The INNOVATORS

The Discoveries, Inventions, and Breakthroughs of Our Time

John Diebold

TRUMAN TALLEY BOOKS
E. P. DUTTON / NEW YORK

Published in the United States by Truman Talley Books ● E. P. Dutton,
a division of Penguin Books USA Inc.,
2 Park Avenue, New York, N.Y. 10016

Published simultaneously in Canada by Fitzhenry and Whiteside,
Limited, Toronto.

Library of Congress Cataloging-in-Publication Data

Diebold, John, 1926–
The innovators: the discoveries, inventions, and
breakthroughs of our time / John Diebold. — 1st ed.
p.   cm.
"Truman Talley books."
Bibliography: p.
Includes index.
ISBN 0-525-24830-7
1. Technological innovations.   I. Title.
T173.8.D54   1990
600—dc20
89-34510
CIP

Designed by Steven N. Stathakis

1 3 5 7 9 10 8 6 4 2

First Edition

Grateful acknowledgment is given for permission to quote from "The News of Radio"
column, *The New York Times*, July 1, 1948. Copyright © 1948 by The New York
Times Company. Reprinted by permission.

For Emma—
It will take a lot
of innovation
to make your world
what I would wish it
to be for you.

Contents

Contents

Preface

This book is a series of true narrative histories, selected from a wide range of innovative individuals and organizations, that examines the many implications of an old word that has undergone a renaissance in recent years. The word has a simple basic definition—"A new idea, method, or device"—but evidence that companies must be innovative in order to prosper in the face of increasing international competition has given it new importance. *The Innovators* described in these pages are those who have already accepted the challenge: dedicated individuals, teams, and organizations that strive to take new ideas, methods, and devices from workbench to marketplace and improve humanity's lot in the process.

Selecting the significant achievements of recent decades has proved to be especially difficult but we have narrowed the field by concentrating on innovations that have arisen within organizational structures (rather than those of lone inventors) during the past forty years or so. It has been necessary to omit a number of major triumphs in order to examine innovation processes in a broad spectrum of disciplines and attempt to learn how innovators produce genuinely positive results within quite different business frameworks.

Innovation comes in many guises, and it sometimes evolves from accidents or other unexpected occurrences. This book features those serendipitous moments as well as the inspiring stories of gifted people who acted as their own guinea pigs, defied accepted theory, and overcame minimal management support to achieve important breakthroughs that brought glory to their names and new opportunities to their companies. *The Innovators* also documents early struggles of now-powerful corporations whose leaders believed in dreams and rode those dreams to enormous success. But innovation requires will as well as skill, and this book also examines companies that became powerful and famous because of innovation—only to fall into decline when prosperity led to a loss of focus and to growth for growth's sake.

If we have succeeded, these pages should come alive with the true stories of individuals from many walks of life who had ideas they refused to abandon, plans they were compelled to carry forward despite considerable odds. Traditionally, most great innovations grow from the unshakable convictions of a single champion, usually because he or she sees a need or an inadequacy in earlier methods or products. But in today's more competitive world, not even the most brilliant innovator can ensure the realization of his vision alone. This book, then, explores the contributions of the many who are necessary if a significant advance is to permeate the society and deliver the humanitarian and economic rewards it was designed to produce. This is what we consider the process of innovation.

Most of all, the stories you are about to read are a tribute to the human spirit. Behind the economic riches, the vigorous corporations, and the new industries these innovations have spawned are the intensely human stories that make fascinating reading. The innovators you will meet here are the technological and business trailblazers who hewed paths through unexplored territory, the brave souls who refused to believe "It can't be done." They performed their wonders in a variety of fields and all have made our world a more interesting, even wondrous, place in which to live. There is much to be learned from their experiences.

JOHN DIEBOLD
August 1989

1

The Transistor:
Birth of the Age
of Electronics

*"Leave the beaten track occasionally and dive into the woods.
You will be certain to find something that you have never
seen before."*
—ALEXANDER GRAHAM BELL

In 1985, New Jersey's Bell Telephone Laboratories became
the first institution ever to receive the American government's National Medal of Technology. Many view Bell Labs
as the preeminent industrial research facility in the world
and point to its long list of distinguished scientific accomplishments as clear proof of that position. Time has now
verified that few inventions deserve a higher place on that
list than the transistor. A study of the struggle that brought
the transistor into existence becomes an exciting story of
innovation in action.

Congressman Dean A. Gallo of New Jersey, in a 1987

speech before the House of Representatives noting the fortieth anniversary of the transistor, claimed it had "changed our lives forever" and had heralded a new era in electronics. "The transistor's switching powers," the congressman stated,

> provided the technology for the computer revolution that has now become an integral part of our lives and made the drudgery of paper recordkeeping and filing obsolete. . . . Because of the transistor's tiny size thousands of transistors and diodes can be woven into a maze of circuitry smaller than a pencil eraser.
>
> Without transistors . . . there would be no pocket calculators, and computers and battery-operated radios and televisions would be much larger and more expensive. Transistors are also the ghosts in such machines as heart-beat regulators; certain cameras; electronic guitars and instruments; hearing aids; electronic watches; and radio, video, stereo systems. Transistors are the cornerstone of space flight and modern aeronautics. These minuscule powerhouses are also the driving force behind satellite communications. . . . Today, it would be hard to imagine what life would be like without this small but extremely powerful piece of technology.

Even allowing for local pride and political hyperbole, it is difficult to argue with the congressman's assessment. Few innovations have had a more widespread impact on our times. Three distinguished scientists—John Bardeen, Walter H. Brattain, and William Shockley—are recognized as the inventors of the transistor and were awarded the Nobel

Prize in Physics in 1957 to honor their outstanding achievement. But, in a very real sense, dozens of scientific innovators in a variety of fields made integral contributions to the project as a normal part of their day-to-day work at AT&T's Bell Labs. The story of the transistor does much to explain the momentous changes in the innovation process since the days of Alexander Graham Bell, the man who founded the sprawling research center that bears his name.

The American Bell Telephone Company was established in 1880, just four years after Bell invented the telephone, but a research branch of the company's engineering department, Bell Laboratories, was not instituted until 1925. It made its presence felt early when Clinton J. Davisson demonstrated the wave nature of matter, just two years later, and became the first of seven Bell scientists to win the Nobel Prize in Physics.

The primary business of Bell Labs has always been to provide AT&T the technology base it needs to be a major leader in the field of information systems and services. Bell engineers had been using vacuum tube technology in telephone systems since 1915, spurring its widespread use in a variety of other practical services. But, by the mid-1930s, Bell Labs started a research program with the objective of introducing electronic switching into the telephone system.

Dr. Mervin Kelly, then director of research, was looking forward to a time when the metal contacts used in exchanges to make connections when numbers are dialed would be replaced by electronic devices. These devices would have to be more reliable, less bulky, and more cost-effective than anything previously used.

The first breakthrough pointing toward such devices came in 1939 from an entirely unexpected source—the Bell radio division's research center in Holmdel, New Jersey. Russel Ohl, a chemist turned radio engineer, was attempting to improve the methods of detecting very-short-wave radio signals. He had discovered that bringing a fine metal wire into point contact with a crystal was the improvement he was seeking, and experimentation convinced him that a silicon crystal suited his ends better than any of the other materials he had tested.

It was difficult, at that time, to obtain uniform silicon samples, and Ohl asked two Bell metallurgists, Henry Theurer and Jack Scaff, to produce some for him in their own laboratory. Scaff and Theurer complied with his request and supplied him with a number of silicon ingots that Ohl sent out to a jewelry manufacturer in Perth Amboy to be ground down and cut up into cylinders about an inch long and an eighth of an inch in diameter.

It was on a hot summer day, several months after he had received them, that Ohl finally got around to measuring the voltage of Scaff and Theurer's cylinders. In an effort to keep cool, Ohl moved an electric fan down to his workbench near the cylinders and his oscilloscope. He noticed some strange reactions on the oscilloscope as soon as he began testing the cylinders. The voltage fluctuated wildly in a somewhat rhythmic pattern.

It took him a very few minutes to determine that the warming beam of his worklight was passing, inadvertently, through the blades of the fan on his workbench. An electrical current was being produced when the light fell on the silicon and, in turn, being drastically reduced when the blades of the fan diminished the light.

Neither Ohl nor any of his associates had ever seen anything comparable or had any idea what caused it. Following standard procedure at Bell, Ohl took his accidental discovery to his immediate superior, who passed it to his superior, and so on up the line.

Among those who witnessed the strange phenomenon soon after its discovery was Walter Brattain, who would come to understand years later exactly what was happening and apply that knowledge to his development of the transistor. At the time he said he was "flabbergasted. I thought my leg was being pulled."

In essence, Russel Ohl had accidentally discovered the photovoltaic effect in silicon as reflected in the crude semiconductor unsuspectingly put together by Jack Scaff and Henry Theurer. In attempting to supply Ohl with uniform ingots of silicon, Scaff and Theurer had "doped" the material with electrically active impurities that produced the voltaic effect when the light hit it.

In the wake of those remarkable accidents, the stage was being set for the ultimate breakaway from the vacuum tube at the beginning of the 1940s. William Shockley's notebook, dated December 29, 1939, proposed a semiconductor amplifier that he still regards as the "first milestone on my path to the concept of the junction transistor." Unfortunately, the theoretical design did not work.

What was working, with deadly precision at the time, was Hitler's war machine in Europe. It caused nearly all of American industry to change course, and most of the resources of Bell Laboratories were soon rechanneled into the war effort. Bell scientists, particularly, were at the forefront in the race to perfect radar. Brattain and Shockley left in 1942 to work in military research programs.

It wasn't until 1945, after Shockley and Brattain had returned, that Bell Labs was able to resume its concentrated research on a semiconductor amplifier. Shockley made several trips to New Jersey from his Pentagon office to review some of the advances made by Bell Labs in semiconductor technology during the war years. Silicon and germanium had become recognized as two of the best controlled semiconductors, thanks to the work of Bell people at Holmdel and

Murray Hill, New Jersey. Mervin J. Kelly, now the executive vice president of the laboratories, had been authorized to make sweeping changes in the overall organization of the company. One of his specific aims was new progress in semiconductors through solid-state physics.

Kelly named Shockley as co-supervisor, with S. O. Morgan, of a solid-state physics group. Shockley almost immediately brought a theoretical physicist named John Bardeen into Bell Labs as a member of the team. Bardeen had spent the war years with the Naval Ordnance Laboratory in Washington, D.C., after earning his Ph.D. at Princeton. He shared an office with Brattain and Gerald Pearson, who had been involved with semiconductor research before the war.

Shockley, with the authority of Kelly and Buckley, had put together an extraordinary team of practical and theoretical physicists whose talents complemented one another. The group's first task was to review everything known about the workings of semiconductors. Shockley, as the result of one of his wartime visits, had already begun to direct his thoughts toward inventing a semiconductor amplifier as a result of the information gleaned by Bell scientists during the war.

Using what he would later describe as the "try simplest cases" approach, Shockley created the first design for a device that might replace Lee De Forest's triode audion in April 1945. Unfortunately, when Brattain and his colleagues applied a strong electrical field perpendicular to the thin slab of semiconductor material, as Shockley had suggested, the modulation of current along the slab did not happen as he had envisioned it.

John Bardeen was assigned the task of analyzing the failure. He confirmed the accuracy of Shockley's calculations but remained equally puzzled by the disparity between theory and experiment. Bardeen took nearly a year to solve the problem of the field-effect failure, and his detailed explanation became one of the most creative and significant research avenues of the semiconductor program. Bardeen's explanation also cleared up a number of other previously perplexing observations about semiconductor surfaces.

Still, another year passed before significant progress would be made in controlling the electron flow in semiconductors. A study of the dated and witnessed Bell Labs notebooks shows that Shockley wrote little about p-n junctions until March 1947. In the meantime, Bardeen and Brattain were continuing their experimentation based on Bardeen's surface-states theory. But activity, and entries in the scientific notebooks, began to increase dramatically in November 1947.

Brattain and Bardeen had enlisted Robert Gibney, a physical chemist Shockley had recruited from the chemistry department shortly after the war, to assist them in improving the performance of the semiconductor. Gibney, as reported in a notebook entry by Brattain on November 17, made the key suggestion that voltage be applied between the metal plate and the semiconductor while both were immersed in an electrolyte. The new approach worked.

As William Shockley would write later,

> This new finding was electrifying. At long last, Brattain and Gibney had overcome the blocking effect of the surface states—the practical problem that had for so long caused the failure of our field-effect experiments. . . . The new experimental conditions set up to study surface states and the resulting new facts about penetration of the field into the surface motivated the will to think of ways to achieve the practical result of a semiconductor amplifier. Within a week, Bardeen, Brattain, and

Gibney had conceived two of the five device inventions that were filed for patenting before the public announcement of the transistor.

That breakthrough of November 17, followed by a twenty-nine-day period of frantic activity leading to the first point-contact amplification on December 16, caused Shockley to name it "the magic month." It was now clear that surface states could be overcome and the resultant thinking and action could be directed toward the creation of a real physical entity. Bell notebook entries reveal that achievements came in rapid order:

> December 4—Walter Brattain reports success of several device ideas including the modulation of p-n junction resistance with voltage applied to a drop of electrolyte over the junction; December 8—William Shockley writes concepts of a junction, field-effect transistor and of voltage gain by using reverse bias on a p-n junction; December 8—Brattain enters achievement of voltage and power gain by using reverse voltages on "high-back voltage" germanium; December 16—Brattain's notebook entry records first voltage and power gain in a point-contact transistor.

A "magic" twenty-nine days, indeed!

In his 1956 Nobel lecture, John Bardeen gave a brief but eloquent description of the key events during the second week of December 1947:

> It was next decided to try a similar arrangement with a block of n-type germanium. . . . [After

some less than satisfactory experiments] We next tried to replace the electrolyte by a metal control electrode insulated from the surface by either a thin oxide layer or by a rectifying contact. A surface was prepared by Gibney by anodizing the surface and then evaporating several gold spots on it . . . a point-contact was placed very close to one of the spots and biased in the reverse direction. A small effect on the reverse current was observed when the spot was biased positively but of the *opposite* direction to that observed with the electrolyte. An increase in positive bias *increased* rather than decreased the reverse current to the point contact. The effect was large enough to give some voltage, but no power amplification. This experiment suggested that holes were flowing into the germanium surface from the gold spot, and that the holes introduced in this way flowed into the point contact to enhance the reverse current. This was the first indication of the transistor effect.

It was estimated that power amplification could be obtained if the metal contacts were spaced at distances of the order of 0.005 cm. In the first attempt, which was successful, contacts were made by evaporating gold on a wedge and then separating the gold at the point of the wedge with a razor blade to make two closely spaced contacts. After further experimentation, it appeared that the easiest way to achieve the desired close separation was to use two appropriately shaped point-contacts placed very close together. Success was achieved in the first trial; the point-contact transistor was born.

On December 23, Brattain and Bardeen invited a select group of colleagues and Bell executives to a secret demonstration of the long-sought device. Everyone was jubilant. The innovative seeds first planted in the mid-1930s were ready to be harvested.

The notation in Brattain's workbook, dated Christmas Eve 1947, seems very low-key today: "The circuit was actually spoken over and by switching the device in and out a distinct gain in speech level could be heard and seen on the scope presentation with no noticeable change in the quality."

There was reason for jubilation inside Bell Laboratories that Christmas season of 1947. Brattain and Bardeen had actually created a device that would become known as the "point-contact" transistor. But precious few people were aware of that knowledge—even inside Bell Labs.

The Bell hierarchy's decision to keep the new invention a secret grew out of a number of factors. Under their terms of employment, Bardeen and Brattain were legally obligated to turn over their patent rights to the company in exchange for $1.00 each. It was standard procedure at the time. Bell Labs employees now automatically relinquish all rights to their discoveries with no extra money changing hands.

The Bell chieftains also knew that other researchers, particularly a group at Purdue University, were closing in on the mysteries that their own physicists had unraveled. It was decided that a minimum of six months would be needed to examine all the ramifications—legal and otherwise—of this potentially momentous discovery. Much time would be needed to refine and develop what was essentially a crude but workable device. Bell's secrecy was also designed to prevent the implementation of even stronger security measures by the U.S. military, which was certain to be interested in the discovery. No outside interference, however well motivated, would be allowed to override the company's own interests so early in the game. The invention had been an internal Bell Labs project, without government aid.

William Shockley was among those who were both pleased and excited that members of the team he had helped put together had created the first working transistor. But he was not without feelings of ambivalence. He knew that a sense of constructive competitiveness was one of the motivational factors at work at Bell Labs, and he felt distinctly disappointed that his own attempts to devise a transistorlike device had not succeeded. In an interview for this book we asked him to explain his feelings, but he preferred to rely on the sentiments he expressed in a 1976 paper for the Institute of Electrical and Electronics Engineers.

"That notebook entry [Brattain's 1947 Christmas Eve report]," Shockley wrote,

> underlines the significance of one source of the motivations of the transistor group. It is, of course, evident that motivations may be based upon a wide variety of factors. Not the least of these, as reflected in Brattain's notebook entry, is based on attitudes about patents. Even without a financial reward, the prestige factor of an authentic demonstration of an inventive contribution can motivate the obtaining of a patent.
>
> This historical note has "creative-failure methodology" as a central theme. I discuss its role in the interactions among the individuals of the transistor group, including the internal competition in which I myself was involved, and its relevance

to human limitations—particularly my own slowness in recognizing key concepts of the junction transistor.

Shockley was determined, in the near-Christmas afterglow of the birth of the point-contact transistor, to transform his own disappointment into triumph for both himself and Bell Labs. On New Year's Eve 1947, in a Chicago hotel room, he began writing nineteen pages of notes and diagrams that would lead to the conception of the junction transistor. Shockley mailed the disclosures back to his co-supervisor, S. O. Morgan, who witnessed them along with John Bardeen. The pages were later cemented into Shockley's work notebook, which he had left behind at Bell Labs.

Shockley would say later that these ideas would reveal "a p-n-p structure almost—but not quite—involving minority carrier injection into a base layer." Once he returned to New Jersey, it was not until January 23 that he could enter into his notebook the actual conception of the junction transistor. It was this theory that eventually led to the transistor we know today. Shockley applied for his patent for the junction transistor on June 26, 1948—six days before the public announcement of the invention of the point-contact transistor.

One of the curious facts of innovation, from the viewpoint of scientist and layman alike, is the need to create a *name* for the new invention. Even after Brattain and Bardeen had built the first working transistor, they and the others who knew about it simply referred to it as "the device." This couldn't continue—for reasons legal, scientific, and commercial. The two innovators were told, in effect, "You invented it—you name it." And this presented an entirely new problem they were not immediately able to solve.

The device they had created was designed to replace Lee De Forest's three-electrode vacuum tube that he had called an *audion*. They knew that name had not survived and hoped to come up with a short, catchy one that would describe the function of their invention and become part of the language. They tried to narrow the field of possibilities with something akin to *varistor* and *thermistor,* names J. A. Becker had used for other semiconductor devices. But they were stumped until they sought outside advice, as Brattain told an interviewer in 1975.

> Bardeen and I were about at the end of our rope, when one day J. R. Pierce walked by my office and I asked him to come in and sit down. I told him about our dilemma, including that we wanted something to fit with varistor and thermistor.
>
> Now Pierce knew that the point-contact device was a dual of the vacuum tube, circuitwise. He mentioned the most important property of the vacuum tube, transconductance, thought a minute about what the dual of this parameter would be and said, "transresistance," and then said "transistor." I said, "Pierce, that is it!"
>
> My late wife, Kate, when she first heard the name, said that it would probably be shortened to *sistor* before long. When the award of the Nobel Prize to Bardeen, Shockley, and me was announced, my father was out in the Bitterroot Mountains of Idaho. The only way my mother could send him a message was by forest reserve telephone to tell Ross Brattain, etc. . . . So her message was "Tell Ross Brattain the transistor won a Nobel Prize." The message my father received was, "Your *sister* won a Nobel Prize!"

Bardeen and Brattain began preparing the first scientific paper on their invention to be published by *Physical Review*, after the editor had been sworn to temporary secrecy. He agreed that none of the paper's contents would be revealed until two weeks after the world was informed of the news in a New York City press conference. The Bell people were convinced the press conference would inspire an enormous level of excitement. It didn't.

DEVELOPMENT: BRINGING THE PRODUCT TO MARKET

Considering the price paid to produce something called a transistor—nearly a decade of painstaking effort and upward of $1 million (1940s dollars)—the Bell press conference seemed, in theatrical parlance, something of a "turkey." The erstwhile reviews expressed mild interest and little enthusiasm.

The press conference, replete with an eight-foot model transistor on wheels, was held at Bell Laboratories' old headquarters at 463 West Street in Manhattan. *The New York Times* saw fit to print the momentous news on July 1, 1948, among a variety of local program notes in a column called "The News of Radio." Other periodicals also gave the announcement short shrift, and many technical journals either ignored it or delayed reporting it in depth until months after the event.

"A device called a transistor," *The New York Times* revealed,

> which has several applications in radio where a vacuum tube ordinarily is employed, was demonstrated for the first time yesterday at Bell

14

Telephone Laboratories. . . . The device was demonstrated in a radio receiver, which contained none of the conventional tubes. It also was shown in a telephone system and in a television unit controlled by a receiver on a lower floor. In each case the transistor was employed as an amplifier, although it is claimed that it also can be used as an oscillator in that it will create and send radio waves.

In the shape of a small metal cylinder about a half-inch long, the transistor contains no vacuum, grid, plate or glass envelope to keep the air away. Its action is instantaneous, there being no warm-up delay since no heat is developed as in a vacuum tube. . . .

That brief report sums up the immediate and general reaction to Bell Labs' announcement. The transistor was viewed as a laboratory curiosity, and, from a practical viewpoint, the one unveiled in New York in mid-1948 was little more than that. It would be a while before the inventors of the transistor could watch it do all the marvelous things they had planned.

Shockley, meanwhile, was more than ever convinced that his theory of a junction transistor was the wiser course to follow. Shockley has since stated that he put anything potentially useful into his June 1948 patent application— "but not so wild as to be ridiculous. . . . I included heavy doping near contacts, heterojunctions with wide energy gaps to increase emitter efficiency, transit time effects for negative resistance, many-layer structures for modulation, etc."

The reality of minority carrier injection was established during the final half of 1948 and, in April 1949, an "existence proof" germanium junction transistor gave power gain using both emitter and collector p-n junctions. But the key to making Shockley's vision a reality was human persistence, prompting an associate to suggest that he call the planned device the *persistor*.

Others at Bell Labs—particularly Gordon Teal and J. B. Little—were working on a process that would improve the manufacture of the semiconductors on which the transistor relied. They decided to work with germanium, because it melts at a lower temperature, to produce a *single* crystal. Using a technique called "crystal pulling" invented in 1917, the two men managed to produce a sizable crystal that was structurally sound enough to make the necessary transistor action possible.

It was clear, early in 1950, over a year and a half after the introduction of the transistor, that the overall process of perfecting and transforming it into a marketable product was going to require unusually cooperative action. A special team was set up, within the overall structure of Bell Labs, to rush the final development of the transistor. Shockley would direct the physics end of research and development, Addison White would be responsible for those attempting to perfect semiconductor materials, and J. A. Morton would head the general development side of the drive.

The achievement of Teal and Little in devising a method for producing structurally sound germanium crystals paved the way for the new kind of transistor that William Shockley was looking for. But another major problem remained—ridding the germanium of unwanted impurities. This was one of the immediate problems facing White when he was officially inducted into the transistor development team.

After months of experimentation, a member of the metallurgy laboratory, William G. Pfann, came up with a system called *zone refining* that left less than 1 part in 10 million of the unwanted impurities in the refined germanium. The process involved a similar but more advanced idea used earlier in the refining of aluminum. Of equal importance, by more or less reversing the process, Pfann was able to inject needed impurities into the germanium.

Shockley's notes reveal that double doping of a melt was used to grow an n-p-n structure to make a grown junction transistor that was demonstrated in April 1950. It performed according to theory but elicited little enthusiasm. And he says a trip to Britain in early 1951 to lecture at a symposium got him thinking about exploiting the junction transistor's potential. A discussion with military men about the instability of proximity fuses in military shells convinced him that the junction transistor was the perfect device to solve the problem. R. L. Wallace supported the theory, provided the base layer could be made sufficiently thin. Later that year he was finally able to obtain a working model of the practical junction transistor he had envisioned. That transistor was built by Bell Labs' Ernest Buehler, under the direction of Morgan Sparks.

But the now "tamed" germanium remained a problem in itself. It was a rare commodity and hardly the ideal substance for mass production. At that time, a little more than thirteen pounds of germanium was produced annually throughout the world. Bell Labs was able to maintain its supply through the germanium dioxide extracted from the refining of zinc. But the cost of even this form of the element, after purification, exceeded the weight price of gold. Each transistor, using only a tiny amount of germanium, would be ten times more expensive than the vacuum tube it had been created to replace.

The entire transistor group knew that silicon would be the ideal semiconductor for the transistor. The problem was finding a way to make it behave as germanium had. By

1952, Teal and Buehler had managed to grow crystals of silicon, but the substance's high melting point made the process difficult. The work continued, and Henry Theurer eventually devised what he called a *floating zone* method of refining silicon's high melting point. And, later in the year, Calvin Fuller invented a manner of doping both silicon and germanium crystals that was a decided improvement on the process discovered earlier by Gordon Teal. Fuller's diffusion process continues to be the fundamental way all microcircuits are made.

The process of developing the transistor was long and extremely difficult. Five years passed from the time Brattain and Bardeen demonstrated the first working transistor on December 23, 1947, until Shockley and the full resources of the Bell Laboratories were able to overcome the myriad problems of producing a transistor that resembled the product in use today.

Although few understood the marvelous potential of the transistor when it was publicly introduced in 1948, interest in it throughout the scientific community grew apace with the passage of time. The innovators of Bell Labs had created the transistor and obtained the patent on it, and many began to think the company was deliberately withholding it from the market. Actually, teams comprising at least one hundred researchers were utilizing every method available, and devising new ones, in frustrating attempts to refine the device and make it reproducible. The Bell innovators were learning a hard truth: In terms of total effort by highly skilled personnel, it is not unusual for several hundred times more man-hours to be required for development than for invention.

This truth is readily apparent when one examines AT&T's actual record in bringing the transistor to market. It wasn't until October 1951 that Western Electric, Bell's manufacturing arm, actually began producing transistors. They were extremely crude devices, at that, and the primary purpose for their invention—inclusion in the telephone system—did not even begin to become a reality until 1952.

The growing hue and cry that Bell was monopolizing the transistor prompted the company to do a rather startling thing in April 1952. Although Bell Laboratories held the patent on the transistor, it offered to reveal all its processes to anyone who was seriously interested in using them. The company would hold a symposium on the subject on an invitation-only basis. The only string attached was an up-front fee of $25,000 per invitation. In a real sense, the invitation cost amounted to something of a licensing fee. Thirty-five interested parties paid the fee and attended the symposium, enabling Bell to recoup a major part of its pre-1948 investment in the discovery of the transistor.

Bell's openness managed to silence most of the critics, for the time being, and the company soon won more praise for generosity by demonstrating that even an industry giant is susceptible to occasional sentimentality. Alexander Graham Bell had evinced an intense, lifelong interest in improving the lot of the deaf and the hard of hearing. In a salute to its founding father, AT&T offered to grant licenses without any fee to companies that would manufacture transistors to be used in hearing aids. Thus, Raytheon became the first company outside the Bell system to use transistors commercially, producing some ten thousand per month early in 1953. At that time, unfortunately, the high cost of transistors made the new hearing aids considerably more expensive than their conventional counterparts.

Although 1953 marked the first commercial use of Bell's transistor, it also brought the loss of one of the key innovators who had been deeply involved in its development. Texas Instruments was one of the companies that had purchased a license from AT&T, and it managed to lure Gordon Teal away from Bell Laboratories. Teal simply carried on his familiar work in a new environment, and, in 1954, he produced the first commercial *silicon* transistor.

William Shockley also left Bell Laboratories in 1954. When we asked the reason for his departure, Shockley said, "Well, partly because lots of people who were involved in this field had gone off and started businesses . . . and this seemed a challenge to get into. And there were some personal things I was not entirely happy about that made me feel like making changes."

Shockley did open his own business, but worldwide recognition came in 1956 when he, Bardeen, and Brattain were awarded the Nobel Prize for their invention of the transistor. Ironically, 1956 was the same year that Bell lost the rights to the device conceived, invented, and developed in its own laboratories. That year the company waived all rights to its 1948 patent under an antitrust consent decree from the federal government.

THE INNOVATIVE PROCESS AT BELL LABS

The story of the transistor illustrates, as well as any, that innovation in the modern era is seldom a one-man accomplishment. Shockley, Bardeen, and Brattain literally invented the transistor, but the input of an entire troupe of gifted innovators inside Bell Labs helped make it a practical reality. And it was the organization and general philosophy at Bell that permitted innovation to flourish.

From its inception, the managers of Bell Laboratories have encouraged individual initiative within a team framework. Even before the transistor, that approach to innovation paid dividends with research that led the way to such applications as the audiometer, high-fidelity and stereophonic sound, sound motion pictures, live television transmission, and electrical digital computers. AT&T takes pride in what it considers to be a combination of centralized and decentralized approaches in research and development activities, with continual interaction among scientists, engineers, and developers.

Dr. Shockley told us that his "creative-failure methodology" during the transistor period at Bell Labs, described eloquently in his paper for the Institute of Electrical and Electronics Engineers, was not a planned policy at the time but rather a description he devised for it after the fact. Is this approach standard practice for scientists today?

I have gotten into this in trying to do science teaching to freshmen and in talking to other people in different fields. The idea is, you may get creative hunches as the result of having bad hunches. One of the main assets of this approach is that it crystallizes relevant concepts in your mind. One fellow I remember talking to was a professor of dramatic arts in San Francisco who mentioned John Barrymore, the actor. Barrymore said, "I love rehearsals. I hate performances." What Barrymore liked was the creative process of trying to do things and finally getting them into an orderly shape.

In science the same sort of thing applies. I've invented an acronym for this I call *ACCOR*. The A stands for finding key attributes. Then you carry out comparison operations among these key attributes and you end up with orderly relations. So, you've got the acronym A-C-C-O-R. When you're floundering around, trying to deal with a problem, the key attributes will finally emerge.

Of his perception of the innovative process on a personal level, Shockley says, "You don't really understand something unless you can see it in several different ways—in the mathematics and also in the diagrams. Then you can see if there's an accurate one-to-one relationship between the two."

Dr. Shockley continues to praise the basic approach of Bell Labs many years after his departure:

> Was the transistor a product of an engineering program focused on a practical goal, or was the transistor a by-product of pure research unsullied by any motivation other than a search for knowledge? What actually went on was a mixture. . . . In assigning our highest priority to the primarily scientific aspects, we chose those related to the problems that blocked our approach to the long-range practical goal—the creation of a semiconductor amplifier, later to be called the transistor. . . . A few other places must be comparable in quality of personnel and continuity of experience, but I believe Bell Laboratories must be very near the top. This continuity in personnel and spirit means that one can count on about the best that can be expected from management considering the fundamental human limitations that exist everywhere. Thus, in a good organization of capable experienced people, an employee can count upon a really high degree of justice. . . .

Collaboration is the operative word in describing the innovation process at Bell Laboratories. Philip W. Anderson, who shared the 1977 Nobel Prize in Physics for his work in understanding the electronic structure of glass and magnetic materials, says the true value of that approach was demonstrated during the quest for the transistor.

The managers at Bell Labs are fond of saying their primary duty lies in knowing what their people are doing rather than in telling them what to do. An aura of complete openness is insisted upon at Bell, based on the age-old theory that two heads are always better than one. To that end, regular open-session meetings are conducted; there everyone discusses his individual work in front of his peers and the directors. Unnecessary secretiveness, at Bell, can lead to negative evaluations. By the same token, company spokesmen claim that extreme care is taken that staff members rather than managers get full credit for their work.

Bell Labs' S. J. Buchsbaum credits two of the organization's top executives with coining the expression that generally sums up the approach to research management. Arno Penzias, another Nobel Prize winner in Physics and the current vice president for research, and Kumar Patel, executive director of the Physics Research Division, characterize the three key factors in the Bell approach as "the three Fs—funding, focus, and freedom."

In 1985, a federal court approved a consent decree that cost AT&T many of its telephone company holdings and forced a general revamping of the entire organization that included the transferral of thousands from Bell Labs. Not surprisingly, the company told its stockholders that Bell Labs' primary mission—creating specific product lines of business and conducting extensive programs of basic research—would continue unaffected. Not enough time has passed to judge the reality of that optimistic projection. The loss of revenue brought on by the breakup is bound to have a long-term effect on funding for research. Company leaders are again under pressure to devise strategies to offset revenue loss, particularly in methods more closely tying the Labs' work to the finished product.

Ian Ross, president of AT&T Bell Laboratories, contends that Bell Labs will be in the vanguard of a second wave of the Information Age. Ross says,

> What characterized the first wave was the stand-alone, mainframe computer. Today, the mainframe computer may not have gone the way of the dinosaur, but it is on the endangered species list. What characterizes the second wave is . . . networking. Networking that will exploit telecommunications systems on a much larger scale than today. Networking that will allow business, industry, and consumers to marshal information across vast distances in any form—voice, data, or video. Above all, networking that will link and utilize the machine intelligence that has become widely distributed in our factories, offices, and homes.

2

Overnight Delivery: Fred Smith's Federal Express

"Failure is an integral part of the innovation process. You've got to be willing to talk about your failures or you'll never have any big successes."

—FREDERICK W. SMITH

It was not uncommon in earlier years for a solitary individual to fulfill a personal dream and grow rich by bringing to market a clever new way of doing something that dramatically improved the methods then used. Innovator-entrepreneurs of that sort are rare today, but a still-young man named Frederick W. Smith clearly fits that mold.

The basic thrust of Smith's innovation—the creation of an expeditious method for moving vital documents and materials from place to place—has been a continuing goal of mankind since the dawn of time. But Smith envisioned a way of mating modern technology with that ancient objective and set out to establish a company that could generate delivery of packages across the vastness of the American continent—*overnight*! In the process, he effectively invented an entire new industry and permanently altered the world's perceptions about the definition of priority delivery.

Frederick W. Smith heads the Federal Express Corporation, a multibillion-dollar company that remains the clear-cut leader in a booming industry. He prefers to concentrate on the challenges of tomorrow rather than dwell on his near-legendary role in the company's beginning. But the record is clear. The concept of overnight air express began as the dream of a young man who had a passion for flying, a Yale undergraduate named Fred Smith.

With a grandfather who had been a Mississippi riverboat captain and a father who had helped build the family fortune by founding Dixie Greyhound Bus Lines, Fred, Jr., may have been genetically destined to make his mark in a career involving transportation. Certainly, airplanes had always fascinated him, and he earned his pilot's license at the age of fourteen. But he had no specific career plans when he entered Yale University in 1962 and majored in economics and political science.

His devotion to aviation inspired him to think about weaving that subject into an economics term paper he was required to submit during his junior year. He had already begun to formulate some strong ideas about the future of aviation while having the time of his life as a standout member of the Yale Flying Club. It seemed to him that business had not yet begun to take advantage of the possibilities of commercial aviation, especially the speed of jet airplanes. Why was it that jet airliners could move *people* from one edge of the continent to another in six hours while an air mail letter from Memphis usually took two days to reach him in New Haven? The system wasn't working the way it should.

Everything he had been taught stressed that modern technology was changing the economic geography of the nation. New industry was springing up all over the country, no longer restricted to sites along major waterways or rail lines or in close proximity to traditional raw materials. If America was moving away from heavy manufacturing into service and high-tech industries headquartered outside the normal traffic lanes, the problem of moving needed materials across vast distances was going to grow more acute.

The big automakers and other manufacturers kept large inventories of parts on line for emergencies, but small companies were not equipped for that. Nor could they afford their own airplanes to transport priority materials as IBM and Xerox did. He had noticed that those companies were already airlifting material out of the small airports near Yale where he did his own flying.

Further research convinced him the idea he was developing for his term paper had practical as well as theoretical merit. If the watchword of the Information Age is "instant access," it was time someone devised a delivery system that could give those service companies speedy logistical support. Smith saw the Federal Reserve System as a perfect example of an organization operating at a nineteenth-century pace because of its inadequate courier system. It was perpetually gridlocked because of the lengthy time gap between the issuance of a check and the actual collection of funds. Only a radically new airborne delivery system could help it cut down on the float. Landing a government agency like the Federal Reserve as a client would ensure an air express company's success.

Flying Tigers and Emery Air Freight had begun flying special delivery goods at the conclusion of World War II, but they had limited access to out-of-the-way places. Air deliveries by the U.S. Postal Service, Railway Express, and others depended on the schedules of passenger airliners, and additional time was wasted trucking packages long distances to and from major airports.

Smith's theoretical delivery company, as explained in his paper, would have a sizable fleet of its own planes, trucks to pick up and deliver to regional airfields, and a central package-rerouting facility located at a major airport near the heart of the country. His trucks would pick up from the customer and carry the packages to the "spoke" airfields from which the jets could fly them to the centrally located "hub." They would be quickly sorted, rerouted, and flown out from there to the spoke airport nearest the final destination, where ground couriers would finish the deliveries. Flying in the light-traffic hours of the night would help assure the customer of overnight delivery.

Fred Smith's enthusiasm for his theory caused him to work overtime, resulting in a late delivery of his term paper. His economics instructor found the idea more utopian than practical. Astronomical financing would be required, and the airlines and others were sure to provide stiff opposition. Federal regulations governing air cargo could also stymie the plan. What grade did Smith receive for that now-historic paper?

"That's one of the myths that has grown up about all this," Smith says today. "I don't have a clue as to what I got on that paper. I was asked by a journalist, repeatedly, what grade I made. I finally said, offhandedly, 'I guess it was my usual gentlemanly C.' "

Fred graduated from Yale in 1966 and, having gone through the ROTC program, received a commission as a second lieutenant in the Marine Corps along with his bachelor's degree. He had hoped for eventual assignment as a member of the Corps' judge-advocate branch, but he was quickly sent to Vietnam to lead an infantry platoon in the thick of the fighting. A year later, with more than two years' active duty remaining, he signed up for Marine pilot training, and that eventually led to reassignment in Vietnam—this time in forward air control.

Smith returned to civilian life determined to make a living from some branch of aviation, the subject he knew and liked best. He joined Arkansas Aviation Sales in Little Rock, a company that was producing annual revenues of about $1 million when he arrived in 1969. He bought a controlling interest and, specializing in the purchase and sale of used corporate jets, upped the annual income to $9 million within two years.

It was this first hands-on experience in business that brought Smith back to the air express idea he had envisioned at Yale. Getting prompt delivery of the parts he needed to refurbish planes before putting them on the market was his biggest problem. Why should fast, dependable delivery be impossible? There must be hundreds of other businesses experiencing similar difficulties. The more he thought about it, the more convinced he became that his five-year-old theory was a basically sound idea whose time had come. Why wait for someone else to do it? He commissioned two consulting firms to make a detailed study of the existing air-freight industry, and their reports verified his own conclusions. There was widespread dissatisfaction with all forms of delivery. Well over half of the scheduled air service was between the twenty-five largest cities in the nation and 80 percent of the smaller, high-priority packages that interested him had to be picked up and/or delivered outside the main traffic corridors.

As for federal regulations, the Civil Aeronautics

Board set very strict standards for the large cargo aircraft used by national air-freight shippers. However, these restrictions would not apply to lighter jets suitable for the shorter runways of small airports. The study gave strength to another of his premises: nine out of ten commercial aircraft were not in service between 10:00 P.M. and 8:00 A.M. That was when his birds could soar unimpeded. He could do it!

To demonstrate his own confidence, Fred Smith invested close to $4 million he had inherited from his father, and Federal Express was incorporated on June 1, 1971, as an air charter company headquartered at Little Rock. Wanting a corporate name that indicated the nationwide scope of the company's service, he reverted to his college thinking and borrowed the first name of the Federal Reserve System, whose business he still hoped to attract. The name proved to be an excellent choice, but the federal agency spurned his bid in favor of developing its own air system.

To his chagrin, Smith was not immediately inundated with the big-number checks he had expected from those who agreed there was a dire need for a nationwide express delivery service. He put his flying skills to good use for the next year, making repeated trips to New York, Chicago, and elsewhere in frustrating attempts to raise capital. Smith adapted the dogged determination that had kept him effective in Vietnam to make himself a dynamic advocate for his new venture and after a year of pleading his case before private investors and banking concerns, he had raised a total of $72 million in loans and equity investment. That figure became one of the largest single venture-capital start-ups in American business history, but it still left little margin for error. His hometown of Memphis, Tennessee, would be the "Superhub" headquarters of Federal Express.

In jet time Memphis was not far from the nation's center, its airport was a modern facility that averaged about ten hours of unflyable weather per year, and it had very little traffic between midnight and six in the morning. The Memphis–Shelby County Airport Authority also agreed to reasonable long-term leases on the property and facilities required, and Smith was confident the area could provide the kind of work force Federal would need.

Federal Express would start on a somewhat smaller scale than originally planned. Package size would be limited to seventy-five pounds, and the company would also offer a lower rate for second-day delivery. As an extra inducement, Federal would deliver "anything that would fit into a manila foolscap envelope for a flat fee of five dollars." It would be an air-taxi service—an airline for small packages only.

Arrangements were made to fly Federal Express planes in and out of thirteen airports serving twenty-two major cities. To expand that coverage a fleet of vans was purchased to rush the packages from clients to airplanes. Smith chose a French-built executive jet, the twin-engined Dassault Falcon, as the first aircraft to bear the Federal Express logo. Stripped of the niceties to make room for cargo, the Falcon could carry 7,500 pounds at maximum speed and minimum expense. Federal placed an order for thirty-three of the Falcons, receiving almost 25 percent off the original price for ordering in quantity.

Smith and his backers opted for a multifaceted approach for heralding the arrival of the new freight company. All packaging, vehicles, and aircraft would flaunt the bold Federal Express logo in bright orange, purple, and white. Personal letters were delivered to desirable corporate clients, and ads in newspapers and distribution trade publications and direct-mail appeals all billboarded the revolutionary new system that guaranteed delivery of priority packages to any destination by noon the following day.

The theme of the campaign was implicit in Smith's own words:

We're a freight service with 550-mile-per-hour delivery trucks. This company is nothing short of being the logistics arm of a whole new society that is building up our economy—a society that isn't built around automobile and steel production, but that is built up instead around service industries and high technological endeavors in electronics and optics and medical science. It is the movement of these support items that Federal Express is all about.

The upbeat advertising campaign was also destined to serve a less obvious purpose, even as the company trained its employees and polished the system through test runs before the official start of business. The corporation's once-mighty pool of financial resources was growing dangerously low, and Fred Smith was hoping his efforts would entice new investment capital as well as customers. Official operations would begin on April 17, 1973.

EARLY TRIALS, TRIBULATIONS, AND TRIUMPHS

It's unlikely any $72 million company ever opened its doors to as little business as Federal Express did on its first day of operation. The vans hummed, the Falcons roared, and the system worked to perfection. But when Fred Smith and his corps of package handlers gathered in the big central-sorting building that night, they were greeted by a trickle instead of a flood. Mr. Smith was able to laugh about the experience when we interviewed him for this book. He was right at the hub and watched as only *eighteen* or so packages arrived. Did he feel that was a sign of the country's profound indifference to his innovation?

"The main thing isn't what I felt," he grinned, "but what I learned. I learned to be cautious in accepting sales people's estimates of volume. We expected to be *buried* in packages that night. I'm sure I was just sick. It was extremely disappointing."

That early disappointment was short-lived. Business began to pick up rapidly, as word spread that Federal Express delivered what it promised, and that enabled the company to expand its service. The daily package count passed a thousand within weeks and continued to climb. Fred Smith's airborne innovation was clearly beginning to demonstrate its potential.

There was no time for reflection at that stage of development, but Fred Smith has given a great deal of thought to the subject of innovation since that time. Asked for his definition of an innovator and the source of that creative individual's inspiration he says:

> I would say that most innovators, or entrepreneurs, have some sort of inner-directed zealotry that comes from some psychological impetus of one sort or another. I mean, take Thomas A. Edison; now, he was an inventor but he was certainly an innovator, too. The man was just totally driven in this vision that he had. I think that innovators just have this view of something and, in their minds, it's extraordinarily important and it just has to be done.

Where it comes from—I don't know. I think in my own case, it's probably having had a father who was successful. . . . I think there was some ideal in that regard. I think some of it probably came from the experience I had in the Vietnam War, being very anxious to do something constructively. It's some sort of inner-directed vision or unfulfilled aspiration that just consumes the individual. As part of the equation, these ideas generally come from some experience or some educational or technical base. In the case of Steven Jobs or Ken Olsen, they did what they did because they got interested in computers. In my case, I was interested in what became Federal Express because I'd been interested in aviation most of my life.

Smith was given a dramatic new reason to worry about the survival of his vision when the OPEC nations precipitated their infamous oil crunch a few months after the corporation opened for business. The start-up costs had drained the company treasury, and outrageous new fuel prices began causing financial problems that could not have been anticipated. Despite a daily package count of three thousand early in 1974, the company was losing *$1 million a month* by the middle of that year.

Sound plans to attract new customers were on the table, but those plans would require additional funding, and the company was sliding into technical default. Could anyone raise new funds for an organization whose investors had spent more than $72 million already for the opportunity of losing an additional million every month? The immediate answer was a resounding no! It was during this period that the determination of Fred Smith and the loyalty of his employees began to assume legendary proportions.

34

To help keep the system operating, Smith sold his personal airplane. Company old-timers tell tales about couriers who left their own watches as security against the purchase of gasoline, and pilots who hid their Falcons when tipped of the imminent arrival of officers from the sheriff's department. Unpaid bills were mounting with the quickness of a jet altimeter, but service was continuing and morale was remarkably high. This made Smith's futile attempts to find new financing even more frustrating.

Fred Smith did something so whimsically offbeat at the conclusion of an unsuccessful fund-raising trip to Chicago that it has since taken on a mystical aura. He was in O'Hare Airport, almost ready to admit defeat on his return to Memphis, when he saw a schedule announcing the immediate departure of a plane for Las Vegas. Many were then calling his innovative approach to air freight little more than a reckless financial gamble, but Fred had never been much tempted by cards, dice, and the other enticements of traditional gambling. Still, on some unexplainable impulse, he bought an airline ticket to Las Vegas.

On arrival in the gambling resort, he sought out a blackjack table and began to play. He was prepared to spend exactly what he had in his pockets—something close to $200. After a bad beginning, his luck changed and he started winning big. At the end of the evening, he had accumulated $27,000 in chips. That wouldn't begin to cover the payroll, but it would help. More important, he saw his good luck as a portent of better things to come. It proved to him that anything was possible. It was no time to surrender.

Fortuitous timing in the form of a disabling labor strike against a competitor also helped to improve the immediate prospects for Smith's ailing company. The United Parcel strike forced that company to cut back on many of its standard services, enabling Federal Express to take up the slack. During the same period, rumors abounded that Railway Express Agency was on the verge of collapse. Smith moved on to the offensive with a method for attracting new customers and was soon able to add another $11 million to keep the company alive.

As Smith and marketing adviser Carl Ally saw it, the growth of Federal Express had been slower than anticipated because the company had not made full contact with the market it had been created to serve. In its first year of operation, company advertising had been confined to the traditional media distributed to and read by only those directly responsible for shipping and delivery. A potentially bigger market was not even aware that direct overnight delivery was possible. These people—the heads of companies, the secretaries responsible for getting urgent material on its way, even those who only wanted to ship a present to a family member at a distant location—had to be reached. Federal Express was going to take advantage of the medium used to sell other widely used products: *television.*

The new TV commercials focused on the company's first fleet of jet airplanes and paraphrased the founder's own words for its catchline: "FEDERAL EXPRESS—A WHOLE NEW AIRLINE FOR PACKAGES ONLY." With a budget of $150,000, the announcements aired in six markets in 1974, and their impact caused the nightly package count in Memphis to jump from three thousand to ten thousand within a short period of time. But Federal was still operating at a loss and lagging far behind the more established carriers in 1975. The company's commercials began drawing direct comparisons between Federal and its competition, and the Superhub was processing eighteen thousand packages nightly by 1976.

Clearly spurred onward by its effective advertising campaign, the company began operating in the black for the first time that year. Federal Express had lost $27 million in its first two fiscal years but managed a remarkable turn-around to produce $75 million in revenues and $3.6 million in profit in fiscal 1976.

The company had continued to expand its services as its financial health improved, but its Falcons were still limited to 7,500-pound cargoes under its license as an air-taxi service. With these restrictions and the ever-escalating price of fuel, Federal Express was nearing a saturation point where further profits would be impossible. The regulations must be changed. If other industries could lobby Congress to pass laws designed to improve the business climate, Federal Express had the same right.

Still without the resources to engage a professional lobbying firm, Smith and his employees embarked on a genuine grass-roots campaign to influence representatives to their way of thinking. Employees across the nation organized a letter-writing campaign to their representatives while Smith made personal pleas to younger legislators and their aides known to be open to progressive ideas. The effort was successful and legislation that cleared the way for the new type of all-cargo air carrier that Fred Smith had pioneered was passed in 1977. The lobby was effective enough that many nicknamed the legislation "the Federal Express Bill."

That law, authorizing the use of larger aircraft, firmly cleared the way for air-express companies to become the force they are today. Federal immediately expanded its capacity many times over by buying used Boeing 727-100s from airlines to complement its fleet of Falcons. The nightly package count, revenue, and profits had more than doubled by the end of fiscal 1977, and, anticipating a growth rate compounding at about 40 percent a year, Federal Express became the dominant force in the industry in 1978. Smith and his associates then decided it was time to take the company public. Priced at $3.00 per share, 1,075,000 shares were offered as of April 1978. Within two years the stock had risen to $24 and it has continued to escalate.

More deregulation allowed Federal to make quantum leaps forward in other areas. The Interstate Commerce Commission ruled that ground services provided by air-express companies would be exempt from regulation and Federal Express won a certificate authorizing it to carry commodities between points in the contiguous forty-eight states. A change in postal rules enabled the company to compete directly with the U.S. Postal Service in the delivery of overnight letters. This ruling became an immediate boon for Federal Express and others and also led to better next-day service and a pricing war that still benefits the general public.

By 1983 the company produced more than *$1 billion* in annual revenue, becoming the first company in American business history to reach that plateau within ten years of its founding. Revenue passed the $3 billion mark in fiscal 1987. The dependable little Falcon jet was phased out in 1984. Just 2 of today's DC-10-30s can carry a combined load of 296,000 pounds, more than all 33 original Falcons could manage together. Today's air fleet encompasses some 21 Douglas DC-10s, 65 Boeing 727s, and 66 small Cessna 208s, giving Federal access to every conceivable type of landing strip and the ability directly to serve about 98 percent of the nation's population. In an era famous for erratic passenger airline scheduling, 98.9 percent of all Federal Express flights in fiscal 1987 were within fifteen minutes of schedule.

Approximately 17,600 vehicles of various types make up the company's current ground-delivery system. Federal Express leases upward of 200 acres of land at the Memphis International Airport and has increased the Superhub's capacity to more than 900,000 packages per night. The relative handful of trained employees of April 1973 has grown to a full-blown army of more than 47,000.

THE PROCESS OF INNOVATION AT FEDERAL EXPRESS

Frederick W. Smith's youthful theory that America was entering a new age where success depended on innovative action and utilization of the sophisticated technology that age was producing continues to be the guiding principle behind every facet of the Federal Express operation.

The first step in trying to perpetuate innovation in an organization is to develop a common set of goals or a common philosophy so that everybody understands what you're trying to do. That way there should be a lot of mental concentration focused on just a few things. What we want here, basically, is to deliver to our customers 100 percent of the time a service that they find satisfactory. Whatever it takes to do that job—we want to do it.

Now, that's a lot different from saying, "Gee, we'd like you to deliver a service that's reasonably okay to the customers and at the same time return a 16 percent ROI and a 12 percent earnings-per-share growth." We have all those goals but they're shared at the very top level of the company so they don't get in the way of the thinking process of the people who are out there actually dealing with the customers. That way they can concentrate on finding a better way to serve that customer. So everybody is very clear about where we want their efforts channeled in terms of innovation.

The second thing we do in trying to foster innovation is to reward it to the extent that we can through monetary awards and recognition. But far more important than rewarding it, we don't kill people for failing. In other words, it's okay to try something—and if it doesn't work—the fact that you *tried* is what's important. Of course, if you've got a repeating record of failures—we don't like that much. But we don't chastise or blame people if things don't work. I just can't tell you how important we feel that is. You've got to allow people the opportunity to fail.

Then the last of the three elements so important to corporate innovation is the need to constantly expose people to new ideas. You've got to get people proficient in the latest thinking in their technical disciplines. You've got to bring in folks who are creative to stimulate their thinking—creating an educational atmosphere, if you will, that keeps people on the leading edge of where you're trying to go, whether that's better station computers, or more reliable radars for the airplanes, or adopting technology for sorting packages faster.

Although completely nonunion, Federal Express offers salaries, benefits, and other incentives that keep it on an equal footing with other companies and has an open-door policy that permits employees to take grievances all the way up the chain of command to Smith himself, if warranted. Federal's vaunted no-layoff policy allowed more than thirteen hundred people, formerly employed in the electronic-mail system, to be moved into other jobs. Encouragement of a familylike atmosphere within the company has led to the naming of each FEC cargo plane after an employee's child, the honor won through a drawing.

Federal Express once employed a single research and development (R&D) group to deal with new technology but now favors decentralized R&D groups within each specialized discipline of the company. "In the final analysis," Smith says, "the people who have to be state-of-the-art are the people who've got to implement as well as dream up the innovations. So you've got to bury that process down in the user groups, in my opinion."

The use of new technology has kept the company at the top of its industry. Federal delivery vans are equipped with radios and on-board computers that the company developed. With its digitally assisted dispatch system, couriers receive accurate and instant routing information on the computer screens inside the vehicles. A newer innovation permits couriers to use a hand-held "super tracker" to transmit information about each package back to Federal's mainframe computers. An anxious client can call Federal's toll-free number and, through the COSMOS IIB tracking system, learn the exact whereabouts of his package within thirty minutes or receive a refund. FEC is also placing on-line computer terminals in clients' offices to integrate billing systems with COSMOS and expects to have ten thousand installed by 1990.

All of Federal's DC-10s and most of its 727s are CAT IIIA–certified, meaning that restricted visibility won't impair their operation. The company created its own weather forecasting system in 1987, further ensuring that its cargoes can arrive on schedule. Federal Express is continually updating its "hub and spokes" system and expanding the capacity of its Superhub. The bulk of all deliveries arrive in Memphis and are sorted, rerouted, and flown out again in a "time window" between midnight and 3:00 A.M., and Federal now has additional sorting facilities on the East and West coasts. Metroplexes at Newark and Oakland airports handle packages shipped between cities within the same region, easing the Superhub's burden by 30 percent.

Obtaining service is easier since Federal split its central call system into fourteen regional telephone centers that can handle nearly a quarter of a million calls every business day. FEC offers discount rates for customers who eschew package pickup service and now has more than 350 Business Service Centers in urban areas in addition to nearly 250 drive-through kiosks for retail clients along major thoroughfares, in shopping centers, and in office parks.

The first step in taking Federal's high-priority logistics system to the international market came in 1984 with the acquisition of Gelco Corporation, an established international courier business, that allowed Federal to begin direct transatlantic service the following year. Irish and British companies were also bought and reorganized, giving Federal 100 percent direct geographic coverage of those two nations.

Federal Express delivers substantial service between the United States and Canada, and company DC-10s fly in and out of a central European hub located at Brussels, Belgium. Directly, through subsidiaries, or by independent contractors, Federal Express picks up and delivers documents and packages in more than eighty-five countries around the world. International expansion has brought on losses, but Fred Smith takes an optimistic view of the future.

> The international business is simply an expression of the fact that we are becoming more and more a global society and, the rhetoric of politicians notwithstanding, they're never going to put the genie back in the bottle. People like to be able to buy

foreign cars and other products from abroad. As the economy has become more global in nature, the same things that created the domestic express business are also taking place in the international business. The only problems are customs barriers, certain protectionist regulations, the great distances involved, and cultural differences. But those four problems can be solved.

As far as the distances—talking about innovation—we're trying very, very hard to get the U.S. government and the manufacturers to build a new generation of supersonic transports that we think can be economically built. One of the things we would have done, if the customs barriers hadn't been so archaic at the time, was to use the Concorde for an international express service to Europe. I gave a speech on innovation and the need for faster transport and that sort of thing some time back and, interestingly, the chief scientist of the Department of Transportation read it and he's come up with a proposal for developing an *unmanned* supersonic international freighter.

Ironically, Federal Express suffered a $223 million after-tax loss in 1987 as a result of its decision to abandon Zapmail, a service instituted in 1984 that appeared to be the ultimate example of Federal's dedication to advanced technology. With the corporation's next-day delivery time already advanced from noon to 10:30 A.M., Zapmail guaranteed *instant or two-hour delivery*.

Federal saw facsimile machines that could transmit copies of documents over telephone lines as a long-term threat to overnight service and invested considerable money developing its own system. Zapmail machines sent instan-

taneous facsimiles of documents and graphics to their counterparts installed by Federal in the offices of heavy-volume
clients. Indirect transmissions were delivered by couriers
with a two-hour guarantee. Federal discontinued the program late in 1986 and phased it out completely the following
May. Fred Smith takes a philosophical view of the costly
failure of Zapmail.

"The point we've tried to get across is—'Look, here
was a real market need. We tried to fulfill it. We did a
reasonable job. We couldn't make the numbers work and
things but, gee, let's put it behind us and go on to the next
thing.' That's what you've got to do if you want to be an
innovator. If you don't have some failures, then you haven't
been trying to innovate."

The man whose vision inspired an entire new industry
remains highly optimistic about the future of the company
that claimed 58 percent of the overnight air-express market
in 1988. He claims that strong competition from United
Parcel Service has only caused the marketplace to expand
much faster than it had in the past. With an eye to the future,
Smith has stated publicly that his company is ready to put
risk capital into a demonstration project leading to the development of hypersonic transports, "In hope that we might
serve as the demonstration arm of this project." He believes
electronic facsimile delivery will continue to impact the business but foresees little likelihood of Federal's reentering that
field except through possible acquisition.

"Bear in mind that about 70 percent of Federal Express's revenues come from moving boxes. That has nothing
to do with electronic transmission. Now, I'll get real worried
when they have one of those 'beam me up, Scottie' devices.
Hopefully, we'll be the first to have one of those!"

3

Cyclosporine: The Transplant Rejection Remedy

"A scientific discovery . . . is never the work of a single person and each of those who collaborated in it has contributed many sleepless nights."

—LOUIS PASTEUR

Hopes raised by the first successful heart transplant operation in 1967 dimmed quickly as that recipient, and the many who followed, survived only briefly—victims of the human immune system, which refuses to accept "foreign" tissue. The innovation that had the power to transform transplant surgery into a viable health-care option was unknown until 1972, when it began to take shape in the laboratories of Sandoz, Ltd., in Basel, Switzerland. It was there that an obscure immunologist named Dr. Jean F. Borel first discovered what would become *cyclosporine,* a drug that safely

45

suppresses the human immune system and allows transplanted organs to survive.

The desperate need for an effective immunosuppressant is illustrated by the history of organ transplantation immediately following Dr. Christiaan Barnard's first electrifying procedure. In 1968, 101 heart transplants were performed in some thirty-six medical centers in sixteen countries. American Dr. Denton Cooley undertook 21 such operations in Texas between 1968 and 1969, but a fatality rate of almost 80 percent caused him to abandon the practice. Another world leader, the Montreal Heart Institute, also gave up on cardiac transplantation in 1970.

Dr. Norman Shumway's group at Stanford University in California was one of the few in the world to achieve increasingly positive results at the beginning of the 1970s, largely because of the sensitive method it developed for detecting early cardiac rejection. But azathioprine and other available immunosuppressants left patients susceptible to anemia, leukopenia and malignancy, and life-threatening infection. Kidney transplants had enjoyed better success, but some safer, more effective immunosuppressant was needed if transplant surgery was to live up to its early promise.

The Pasteur quote that begins this chapter is often used by Jean Borel to soften laudatory claims that he is a "hero" for his personal role in the discovery of cyclosporine. Sandoz, Ltd., has been a longtime leader in pharmaceutical innovations, and dedicated work by dozens of Sandoz scientists went into the effort that made cyclosporine a commercial entity. The nature of new drug discovery requires collective effort, but the record proves that Dr. Borel's individual initiative and determination—even to the point of serving as a "human guinea pig"—provided the spark needed for cyclosporine's success.

Cyclosporine was extracted from a previously un-known fungus retrieved from a rock-strewn plateau in Nor-way, and it came into being because Sandoz began screening fungi for properties other than antimicrobial activity in 1958. The success of the drug azathioprine in the treatment of a kidney transplant patient in the mid-1960s fanned the com-pany's initial interest in finding an immunosuppressant agent of its own. Sandoz established an immunology department in Basel within its tumor chemotherapy group in the phar-macological department, still not looking for specific com-pounds with immunosuppressive activity but hoping such activity might exist in the antitumor preparations being tested.

But the immunology program was steadily enlarged as transplant operations increased. That led to the discovery of a fungal metabolite called *ovalicin,* which was isolated from the broth of the fungus *Pseudeurotium ovalis.* Ovalicin was shown to inhibit production of antibodies, extend the survival of skin grafts in mice, and prevent the symptoms of experimental allergic encephalomyelitis in rats. It also became the first chemically defined nonsteroidal product that demonstrated immunosuppressive activity and no leu-kopenic potential. Hopes were high, but clinical testing proved that it induced more dangerous side effects in hu-mans and the program was abandoned.

The ovalicin research brought improved laboratory techniques, and, in evaluating the metabolites of fungi pri-marily for anticancer activity, researchers discovered that some of them had immunosuppressive potential. The ex-panded screening program led to the recovery of the specific soil samples containing the Fungi Imperfecti from which cyclosporine would come. As part of its early screening pro-gram, Sandoz relied on its scientists to bring home likely looking soil samples whenever they traveled to out-of-the-way places. A Sandoz microbiologist, vacationing in Hardanger, Norway, became infatuated with the stark, primeval landscape of the nearby Hardanger Vidda, a vast highland plateau. He loved the treeless beauty of the place with its windswept grasslands, rocks, lakes, and streams and dutifully collected samples of its soil—that produced mosses, lichens, and occasional ancient arrowheads—and took them back to Switzerland.

The arrival of the Hardanger Vidda soil samples at Sandoz nearly coincided with that of Dr. Borel, who joined the company in 1970 as director of its immunology laboratory. A Swiss national born in Antwerp, Belgium, in 1933, Borel had received undergraduate degrees in engineering and agronomy and a doctorate in immunogenetics from the Swiss Federal Institute of Technology in Zürich. He had spent two years at the central laboratory of the Swiss Red Cross and five more with the medical division of the Swiss Research Institute before joining Sandoz.

In retrospect, there's a degree of irony in the timing of Borel's arrival. Sandoz opted to place less emphasis on cancer chemotherapy shortly afterward and began testing crude microbial broths only for their antibacterial and antifungal effects. This decision pretty much coincided with the waning of international excitement about organ transplant surgery because of an inability to control the rejection factor. Although by then involved in isolating metabolites of microbial origin for screening as antibiotics, Borel had been greatly interested in the findings and the methodology used by Sandoz scientists Hartmann Stahelin, Hans-Peter Sigg, Sandor Lazary, and Hans-Peter Weber during their research into the pharmacological activities of ovalicin. They had learned that such compounds often showed cytostatic or other pharmacological activities greater than their antimicrobial possibilities. It had been agreed to keep such compounds in the screen because they often had unusual chemical structures that made them potentially valuable for new chemical leads. That decision eventually brought the fungus extracted from Norway's Hardanger Vidda and a similar one from Wisconsin into Borel's hands.

The Sandoz Microbiology Department managed to isolate two strains of Fungi Imperfecti, previously unknown, from the Wisconsin and Hardanger Vidda soil samples in 1970. The Wisconsin strain would grow only in surface culture, and it was abandoned. The Norwegian strain, *Tolypocladium flatum* Gams., could be grown in submerged culture, and it was this strain that eventually produced cyclosporine. Zoltan Kis headed a team that isolated a mixture of metabolites on a micropreparative scale from the original crude fungal extract to retrieve what was labeled as novel neutral polypeptides. Sufficient quantities of a two-component mixture, code named 24-556, were produced in the fermentation process to permit biological screening.

In vitro testing demonstrated that the 24-556 metabolite held no great promise as an effective antibiotic. In vivo testing proved that it was just marginally effective, and only then against clinically irrevelant organisms. What did interest the scientists in the animal testing was the fact that the fungistatic activity was coupled with an unusually low toxicity. It was this property that encouraged them to reserve 24-556 and other such metabolites that would enable Jean Borel and Zoltan Kis to conduct further testing in a limited pharmacological screening.

Impressed as he was by the earlier experimentation with ovalicin, Borel was particularly curious about the low toxicity of the mysterious 24-556. Would it have an effect on cancer cells, and was it possible that significant immunosuppressive activity might be present? He had carefully studied the early results obtained with ovalicin and a number of immunosuppressive, cytostatic, and other reference drugs and, soon after joining the company, had modified a mouse test model developed by Stahelin and Lazary. Another Sandoz innovation was the steady development of other in vitro and in vivo experimental models that enabled the researchers to investigate further active compounds in circumstances that resembled, as closely as possible, human testing.

Sandoz began its screening of the Hardanger Vidda soil samples in 1970, and it wasn't until January 1972 that Borel made his momentous discovery. His tests demonstrated that metabolite 24-556 had not only an immunosuppressive effect, but a most unusual one. Using laboratory mice, a series of Borel experiments showed that 24-556 was capable of impressive immunosuppressive activity and, in an even more startling development, its immunosuppression was not linked with general cytostatic activity! This was something entirely new, unrelated to the activity of any other known immunosuppressive agent. Further experimentation, using different size animals, doses, and methods of administering the extremely water-insoluble compound, yielded varying results and frustrations that Borel would not be able to define fully for a number of years.

"It was so beautiful," he told us, "I could hardly believe it myself!" But he kept at it and, by the end of 1973, firmly established the immunosuppressive effect of the compound on both humoral and cell-mediated immunity, as well as its strong activity in Freund's adjuvant arthritis.

What Borel's animal testing seemed to have demonstrated in 1973 was that metabolite 24-556 was capable of suppressing production of certain human *lymphocytes*: cells that protect the body from all manner of foreign invaders including transplanted organs. There are many mysteries yet to be solved about the total working of the human immune system, but it was known at the time that it deploys two types of white blood cells to attack foreign objects that enter the body: The B-cells are the first to be alerted, producing proteins called antibodies to destroy foreign antigens of all types that may lead to infection. If the outside object gains a foothold, two kinds of T-cells begin to multiply: the "helper" T-cells appear first to activate the "killer" T-cells that directly attack—in the case of transplanted tissue—the new organs.

Borel's discovery could be considered little more than preliminary evidence without years of further testing, but it held great promise for organ transplantation. The agents in the 24-556 compound appeared to interfere with the biological communications between the helper and killer T-cells, thus blocking their effectiveness without harming the infection-fighting B-cells. No other immunosuppressive agent worked in the selective manner demonstrated in the simple testing. All other immunosuppressants indiscriminately destroyed rapidly dividing cells, even the vital cells of bone marrow, in the process of preventing cell-related immunity, and that left patients easy prey for dangerous infections and even cancer. Steroids killed T-cells, but long-term use was capable of leaving patients with many other serious afflictions. Borel was convinced he was on the threshold of a discovery of immense import to the entire field of immunology.

Respect for transplant operations had fallen to its lowest level at about the same time that Borel was making his momentous discoveries about the potential of metabolite 24-556. For this and other reasons—the enormous cost of new drug discovery and the need to realize a healthy return on its research investment—Sandoz was forced to reevaluate the overall goals of its research program. Immunological research was destined to be one of the fields most affected. The company felt its involvement with immunology would be better served by integrating it into another major field of research.

The proposed reorganization threatened to end Borel's promising investigation, and this unthinkable possibility compelled him to become the personal champion of the unique immunosuppressant he had discovered. Borel appreciated his employers' reluctance to pour more money into a highly risky venture instead of spending it on programs that were closer to fruition. But Borel's reasons for not shelving the compound were equally compelling. In the end, he was allowed to proceed. He has written: "the management was flexible enough to recognize firstly the strong and reproducible immunosuppressive activity, and secondly the remarkable lack of side effects of this compound as compared with the other reference drugs in clinical use."

THE INNOVATIVE DEVELOPMENT OF CYCLOSPORINE

After winning permission to continue his research, Borel's first task was a deeper exploration of the possible cytotoxic effects of the compound. No biological progress would have been possible without equal advances on the chemical side. The first such advance came with the creation of a system that permitted the purification of cyclosporin A from the mixture that had been called 24-556, which also revealed another form called cyclosporin B. At the same time, others were substantially improving the culture conditions for producing the metabolites, and the two advances permitted the start of pharmacological work using pure CyA (cyclosporin A) in February 1974. The following year, other Sandoz innovators were able to elucidate the structure of cyclosporin A by both chemical degradation and X-ray analysis.

He had won what amounted to a temporary reprieve from his employers, and Borel knew he would have to establish firm evidence quickly to support his enthusiasm about cyclosporin A. With the metabolite purified and larger quantities of it available, Borel began testing to determine its specific suppressive effect on murine spleen cells and mastocytoma cells as well as on bone marrow cell counts and stem cell proliferation in mice. In the in vitro spleen and mastocytoma tests, he conducted identical experiments using azathioprine, hydrocortisone, nitrogen mustard, antilymphocytic serum (ALS), cytosine arabinoside, and colchicine—other immunosuppressants—to analyze and compare meticulously the results obtained with cyclosporin A. He conducted side-by-side testing using azathioprine and cyclosporin A in the studies on bone marrow cell counts and stem cell proliferation in mice.

Borel admits that it was not until this exhaustive study was completed and the comparisons analyzed that he was able to comprehend the full potential of the immunosuppressant he had found. His concept of the selective action of CyA on lymphoid cells was fully supported by the results of the detailed study. It performed nearly every action he had anticipated in a manner that was safer and more complex than he could ever have envisioned.

It had been a bit more than four years since Borel first discovered the marked immunosuppressive effect of the metabolite 24-556 and more than two since the company had permitted him to continue his research. His experiments could have, like so many others in biomedical laboratories, ended in a dead end. But, armed with strong new evidence, he made another direct appeal that the innovative nature of cyclosporin A be recognized and the research continued. This time his superiors readily agreed and promoted cyclosporine to clinical phase A in April 1976.

Shortly afterward, Borel presented his first paper on the results of his animal trials before the spring meeting of the British Society of Immunology. This initial detailing of the unprecedented results from his testing, Borel concedes, probably invited skepticism from many immunologists, but the announcement also generated rapid and unexpected support. Cambridge researcher David White, a colleague of the famed transplant surgeon Sir Roy Calne, said he and Calne would like permission to use cyclosporin A for the first time outside the Sandoz laboratory for heterotopic heart allografts in the rat. The results from that experiment so impressed Dr. Calne that he began using cyclosporin A in a series of transplantation models, and his independent praise for its effectiveness increased interest worldwide.

Subsequent 1975 toxicity tests at Sandoz produced more positive findings, confirmed by respected laboratories elsewhere. Planning began, early in 1977, for the ultimate trial: clinical testing. Borel and associate Dorothee Wiesinger had arranged a straightforward bioassay simply to study the pharmacodynamics of cyclosporin A. In order to measure blood levels, gelatin capsules containing a powdered form of the drug were given to healthy volunteers. But when technicians analyzed blood samples taken over a twenty-four-hour period, they were startled to find absolutely no evidence of cyclosporine's presence. It appeared that the supposedly potent drug had made its way through the digestive tract without ever reaching the bloodstream. If cyclosporin A could not be absorbed in human blood, it was *worthless*!

Faced with the summary cancellation of his long and costly program, Jean Borel stoutly refused to believe the truth of what he had witnessed. It was a deception, he felt, only a technical problem caused by the drug's insolubility in water. After years of observing cyclosporine's activity in animal trials—its easy absorption in the blood of rats and rhesus monkeys—he believed the method of transport was at fault. The gelatin capsule was interfering with the drug's ability to reach the blood.

It was a moment of high drama: the moment of truth for cyclosporine. As Borel describes it, in general terms: "They wanted to give it up, to say now it's finished. You have fooled around enough with this substance and it is costing us, and so on. Oh, yes, it was going to die—to be killed for the last time."

He asked permission to conduct an official experiment on himself, with all the departments involved and the substance controlled to prove the scientific correctness of his theory. He knew an animal could not be killed with one dose of cyclosporine and he intended to take the drug in much the same way, in his opinion an excellent argument in favor of animal testing. Permission was granted. He mixed himself a "cocktail"—500 milligrams (mg) of cyclosporin A dissolved in 160 milliliters (ml) of an unpalatable liquid composed of 95 percent ethanol, 3 percent Tween-80 (a detergent), and a dash of water—and drank it down!

"I got a bit tipsy," Borel admits. More important, he proved his point. Two hours after drinking his experimental libation, using two different bioassays, a significant level of the drug was measured in his blood, proving that it could be absorbed in pharmacologically active concentrations. Heribert Wagner, a Sandoz colleague who had worked on lipid absorption in animals, suggested dissolving the compound in pure olive oil, and that solution proved to be both more tolerable and effective than Borel's self-experimentation. Eventually, other forms of oral administration and injections were perfected, and plasma levels can now be

THE INNOVATORS

accurately determined by a variety of improved methods.

Britain supplied the first surgeons to use cyclosporin A as an antirejection drug in human transplant surgery in June 1978. Dr. Roy Y. Calne of Addenbrook's Hospital in Cambridge, an early booster, prescribed it for use after an operation that transferred the healthy kidney of an unrelated cadaver to a patient. And Dr. Ray Powles of Sutton's Royal Marsden Hospital used the drug for a bone marrow transplant. Both operations were deemed successful, and Sandoz had its first proof that cyclosporin A could exert strong immunosuppressive effects in man.

The early clinical testing of cyclosporin A brought generally positive results, along with some problems that kept dozens of innovative Sandoz chemists, biologists, biochemists, pharmacologists, toxicologists, medical specialists, pharmacists, analysts, engineers, statisticians, and others working overtime. More information was needed about the way the drug worked, and developing the ability to produce larger quantities of it was also a clear priority. The biggest problem was the inability to determine the most effective dosage for patients, brought on by the absence of adequate serum drug assays. Further complications arose from the predilection of many surgeons, in the first clinical trials, to use the potent drug in combination with one of the earlier immunosuppressive agents.

The strong immunosuppressive activity of cyclosporin A in a variety of transplant procedures had been demonstrated, but no fair test of its safety and effectiveness could be made without bold new input from the laboratories of Sandoz, Ltd. Meticulous investigations into every facet of the drug from its chemistry to the formulation of its dosage form to bioanalytical considerations were put into high gear. In 1980 Sandoz chemist Roland Wenger and his team managed to synthesize the complete molecule.

SUCCESSFUL TESTING, APPROVAL—AND QUESTIONS

The ability of Sandoz, Ltd., to provide supplies of the pure cyclosporine for clinical testing at the world's major medical centers became its strongest marketing tool. The drug quickly proved to be its own best salesman. Sandoz was given its first U.S. patent for cyclosporine in 1979 and the federal Food and Drug Administration (FDA) gave it a 1A rating, the highest priority medical advance, with its approval for clinical testing. If the experimental drug's capacity for selectively suppressing the body's natural immune system could be demonstrated clinically, the field of transplant surgery would be revitalized.

Uncovering the best method of using cyclosporine continued to be a problem, with many surgeons' taking a cautious approach because of worries that the drug's potent effect on T-lymphocytes could lead to lymphoma, a form of cancer. Others were concerned about its potential for hepatic and renal toxicity. Efforts to solve these and related problems continued on all fronts, especially at the Sandoz laboratories and at medical centers in Britain and America. Dr. Borel credits two pioneer transplant surgeons, Sir Roy Calne of Britain and Dr. Thomas Starzl of the United States, for key discoveries that eased these worries.

Dr. Calne had visited Basel earlier to urge Sandoz to continue cyclosporine research and had initiated use of the drug in both animal and human trials. He was the first to reveal cyclosporine's tendency to induce nephrotoxicity in kidney transplants but admitted that excess dosage was a more likely villain than the drug itself. He had attained excellent results when he used cyclosporine on dogs, but in humans the same dosage induced nephrotoxic effects that had not been evident in animals. By lowering the dosage for humans, Calne was able to reduce the nephrotoxicity to safer levels. That led to his discovery that lowered levels of the drug reduced the incidence of lymphomas.

Dr. Starzl, who had written the classic textbook on kidney transplants, had been closely monitoring Calne's early use of cyclosporine and was eagerly awaiting his opportunity to use it in the United States. When the opportunity arose, he discovered that the lymphomas that occurred because of cyclosporine's robust immunosuppressive activity should not be removed by surgery. He found that such lymphomas would soon "melt" by themselves if the cyclosporine dosage were steadily reduced. Starzl had originally developed the innovation of combining antirejection agents for optimum results in kidney transplants, and he revamped that general theory to gain more control over cyclosporine's negative potency. He began using controlled amounts of steroids with CyA therapy at the University of Pittsburgh with gratifying results.

Dr. David L. Winter, director of medical research at Sandoz Pharmaceuticals in East Hanover, New Jersey, headed the clinical testing program in the United States. The University of Minnesota Medical School and the Stanford University Medical School were especially instrumental in early demonstrations of cyclosporine's effectiveness as an immunosuppressant in heart transplant surgery. Dr. John Najarian, chief of surgery at Minnesota, claims his was the first American center to use the new drug in 1979 and characterizes its impact as immediate and dramatic. Survival rates for his transplant patients showed an immediate increase of 10 to 15 percent.

Stanford's Norman Shumway, the pioneer who had been expected to precede Dr. Barnard in performing the first human heart transplant, had continued to perfect the technique until he was demonstrating a one-year survival rate of 63 percent even before he began using cyclosporine at the end of 1980. Some 10 percent of Stanford's heart transplant patients endured their initial hospitalizations without any rejection episodes before cyclosporine was used. That rate jumped to a full 40 percent in the first year of cyclosporine therapy. The positive results noted in just a few patients prompted the Stanford team to begin relying exclusively on it as the primary immunosuppressive drug. In less than three years, cyclosporine increased the one-year survival rate to 81 percent.

Clinical testing of cyclosporine in America allowed significant improvements in other vital transplant procedures. Kidneys were first successfully transplanted in the 1950s, and improving techniques kept these procedures viable through the downturn of the 1970s. After cyclosporine was introduced, survival rates for kidney transplants rapidly increased from 53 to 83 percent to 73 to 90 percent.

The advent of cyclosporine also allowed Dr. Starzl to move closer to one of his long-range goals: the perfecting of the extremely difficult liver transplantation in humans. Starzl had performed the first surgically successful liver transplant in 1963 when he was with the University of Colorado Medical Center but discovered that rejection complications were especially acute in that procedure. Liver transplantation had been considered an almost hopeless venture, with a one-year survival rate of only 32 percent. Cyclosporine therapy and surgical skill dramatically changed that, with the survival rate increasing to 54 to 71 percent.

Dramatically improved results with cyclosporine in the three most common operations also inspired even bolder surgical ventures—single-lung, heart-lung, pancreas, and bone marrow transplants. The results were encouraging, and the new drug was especially praised for seeming to solve the problem of tracheal healing in the rare heart-lung transplant.

This inspired a member of the Stanford heart team to predict that cyclosporine's effectiveness, and further advances in surgical techniques, might make heart transplantation as commonplace in the 1990s as the most complicated heart valve replacement surgery of the early 1980s.

It had been demonstrated, by 1983, that the serious side effects evident in early testing could be prevented by meticulous drug therapy: administering CyA with adrenocorticosteroids and not with other immunosuppressive agents. But Sandoz's Dr. Winter cautioned even then that cyclosporine was a potent drug that should not be viewed as a panacea. Winter revealed that the most serious side effects noted in clinical trials were the 27 cases of lymphomas reported from among the 5,400 patients treated—a rate of 0.5 percent compared to the 1.2 to 1.4 percent recorded in treatment with earlier immunosuppressants. Cyclosporine could be mildly damaging to kidney and liver cells, but adverse effects could be controlled by lowering the dose. Increased growth of body hair appeared in some cases and slight hand tremors were evident in others—depending on dosage—and a few patients demonstrated abnormal growth of the gums.

Overall problems recorded during cyclosporine's American clinical trials were considered relatively minor—compared to the monumental risks involved before the drug was discovered. In September 1983, the U.S. FDA gave its formal approval for the marketing of cyclosporine, as soon as Sandoz developed "appropriate" label instructions for physicians. The FDA approval did little to relieve a minor irritant that had plagued Sandoz for years: what to call Dr. Borel's revolutionary rejection remedy. It had been known by different names in different places—cyclosporin A, cyclosporin, ciclosporin, and cyclosporine—generic labels preferred by such groups as the World Health Organization, the British Pharmacopoeia Commission, and the United States Adopted Name Council. Generic spelling differences remain, but the North American trade name is now *Sandimmune*. Everywhere else in the world, it is spelled *Sandimmun*.

Events since the FDA approval of Sandimmune leave no doubt that the drug has had a profound effect on the medical community. In the United States alone, in the first two years after FDA approval, the number of liver transplants doubled while heart and pancreas transplants tripled. Close to nine thousand kidneys were transplanted in 1986 alone, and the survival rates for all transplant operations have continued to increase as more is learned about how to use the drug. Transplant centers in the United States nearly tripled in number during the first three years of Sandimmune's existence.

The innovative work of Dr. Jean Borel and his colleagues at Sandoz in developing a miraculous drug from an unknown soil fungus found on an isolated Norwegian plateau is already viewed as a scientific milestone of the twentieth century. The drug's ability to suppress the body's immune system selectively is preserving life for thousands. Dr. Thomas Starzl told a 1983 congressional hearing that cyclosporine results were so positive at the University of Pittsburgh that "liver transplantation is now considered a service as opposed to an experimental procedure."

The ultimate acceptance of that opinion was demonstrated when private health insurers, and even Medicare, began covering many costs in transplant operations. Cyclosporine's impact on all types of such surgery has stimulated an economic boom for Sandoz and the entire health-care industry. Annual cost of the drug averages about $4,000 per patient, although the pharmaceutical firm says these costs are being reduced as therapy techniques improve. The

company also claims Sandimmune patients benefit economically from shorter, less expensive hospital stays.

But the cost of the drug that has made organ transplants a viable surgical procedure is not the most disturbing
problem its arrival has precipitated. The overriding, overwhelming, and as yet unanswered question is, *who* has the
moral, ethical, and scientific authority to decide which dying
patient has the greater right to a prolonged life through a
transplant operation? Like Goldilocks pondering the por-
ridge of the Three Bears, skilled surgeons must now decide
whether the prospective recipient is too old, too young, or
"just right." Finally, the skill required to perform such a
delicate and complicated operation makes it an incredibly
expensive procedure. Must the ability to pay be the ultimate
factor determining who lives and who must die?

The success of transplant surgery has inspired more
questions than answers, and even the most dedicated advocate of free enterprise must admit that the supply-and-
demand ideal has been of little help. There are simply too
few suitable cadaver organs available to meet a growing
demand. General campaigns to educate bereaved relatives
about the desperate need have had limited success; predeath
agreements authorized by individuals willing to donate their
transplantable organs would seem to be the ideal solution.
Regional voluntary organ banks have helped a bit, and a
number of states now have laws requiring doctors and nurses
to make direct requests for organ donations from the families
of terminal patients.

The most promising option to date was established
by the U.S. Department of Health and Human Services in
1987. The federal agency opened a new computerized center
in Richmond, Virginia, that is designed to serve as a national
clearinghouse for donor organs with computerized services
that match tissue and blood types of available organs to

needy recipients. The new center is intended to make the process more equitable for all, but critics feel this bureaucratic approach will take away the human element of the equation without solving the overriding problems. The moral and ethical questions have no universally accepted answers.

CYCLOSPORINE'S INNOVATIVE FUTURE

To the Sandoz scientists in Switzerland and the United States, the investigation into the innovative uses of cyclosporine is far from concluded. They believe its effectiveness in organ transplantation is only the first established demonstration of its value. In an interview for this book, Dr. Jean Borel told us that he and his colleagues still don't know everything about the manner in which cyclosporine works.

> We still have unknowns, because the basic scientists know not enough about the activation steps of the T-cell. What we know is that cyclosporine inhibits certain of the steps which are known, but if this phenomenon is a consequence of other steps that we don't know—that is the mystery. This is what makes cyclosporine such an important scientific probe, for learning a lot about the activation of the lymphocytes, of the different events taking place. It's also useful to study prophecies, to evaluate the role of T-cells because you can specifically deplete the lymphocytes with cyclosporine and see what happens if you turn them off. The compound is very important from the point of view of basic research, besides being a drug.

The possibility that Sandimmune might become a valuable treatment against autoimmune diseases—afflictions in which the body's immune system turns against itself—occurred to Borel and others a number of years ago. Scientists say the rationale for the use of cyclosporine in autoimmune diseases is nothing less than compelling. Enthusiastic testing has been under way in medical centers around the world to determine methods by which the drug might be used against such diseases. French doctors reported a brief period of success in using cyclosporine against acquired immune deficiency syndrome (AIDS), but their optimism proved to be unfounded.

Borel is convinced that cyclosporine's use in autoimmune diseases will eventually be of more widespread importance than its role in organ transplantation. It has already demonstrated promising results in a number of such diseases. Cyclosporine is being clinically tested in the United States for controlling severe uveitis, which causes inflammation and deteriorating vision in the human eye, and has already been licensed for treatment of the disease in Japan. It is proving to be both effective and beneficial in the treatment of psoriasis, rheumatoid arthritis, myasthenia gravis, and a wide number of other diseases. It has also shown promise as a therapy following skin grafts.

Knowledge gained by cyclosporine therapy in the treatment of type 1, or juvenile-onset, diabetes supports Borel's belief in the compound's value as both research tool and drug. Sandoz sponsorship of cyclosporine trials with this disease began long before a specific connection between it and the immune system had been established. By 1987 there were indications that the immune systems of juvenile diabetes victims were attacking the insulin-producing islet cells as if they were foreign intruders, and cyclosporine was able to protect these cells from destruction in increasing numbers of cases. The drug was stopping islet cell destruction in 60 to 80 percent of the patients undergoing the therapy by 1987, compared to only 30 to 40 percent of those treated when the University of Western Ontario began its program in 1982.

Adults who develop diabetes in middle age or beyond are often able to avoid the need for insulin treatment. The younger victims of juvenile diabetes must receive outside insulin supplies for the rest of their lives. Even then, insulin is not a cure, and patients still face the possibility of blindness, amputation of feet and legs because of circulatory impairment, and fear of life-shortening heart attacks, strokes, and kidney failure.

Cyclosporine therapy for type 1 diabetes is, according to Borel, a special case. He feels the drug has demonstrated its effectiveness, but it must still be determined whether it is truly beneficial. Juvenile diabetes must be diagnosed very quickly to enable the drug to begin defending the remaining islet cells that will allow the patients to depend on their own biological insulin supply. Cyclosporine therapy, like insulin therapy, must be continued for life, and the drug is considerably more expensive. Not enough time has elapsed to prove that Sandimmune is as beneficial, in the long run, as it is effective. Close to half a million Americans stand to benefit if trial results continue to be positive.

Jean Borel, often given the laudatory title "Father of Cyclosporine," has never stopped his investigation of the wondrous drug he discovered. He continues to collaborate with a number of laboratories around the world on highly experimental uses for it. Now the head of his own laboratory at Sandoz, he is leading research on new compounds and feels that cyclosporine has made even more miraculous biomedical innovations possible in the future.

I would say that cyclosporine has opened a new era in what I call immunopharmacology. It has allowed us to learn much more specific things than was hitherto possible. It is the prototype of a new generation and I am sure there will be more compounds—better ones. It may take one month or ten years, who knows, but I know a lot of firms are looking for it because it's a big market, and we will find something.

4

Scotchgard: The Fluorochemical Fabric Protector

"You must kiss a lot of frogs to find a prince . . . the captain bites his tongue until it bleeds!"

—3M FOLKLORE

High on the list of all who spend time studying companies that nurture and support innovators is the Minnesota Mining and Manufacturing Company. 3M, as the company prefers to call itself today, is based in St. Paul, Minnesota, and produces forty-five major product lines in more than forty divisions around the world, a dazzling display of successful variety and versatility. Adding interest to the 3M story is the legend that many of its best-known products, including Scotchgard fabric protector, were derived from "accidental discoveries" in the lab. A study of the thinking that led to Scotchgard sheds much light on this northland behemoth's vaunted strategies for stimulating initiative and creating a climate for innovation.

Scotchgard was commercially introduced by 3M in the mid-1950s, and its family of protective products are now marketed in more than forty countries. The discovery of the soil-resistant properties of the substance was, as legend has it, a complete accident. But the philosophy and process that made it possible for the accident to happen—and be transformed into a lucrative business—were the result of 3M's basic rule of management, laid out by the company's president, William L. McKnight, in 1944.

McKnight said:

> As our business grows, it becomes increasingly necessary to delegate responsibility and to encourage men and women to exercise their initiative. This requires considerable tolerance. Those men and women to whom we delegate authority and responsibility, if they are good people, are going to want to do their jobs in their own way.
>
> Mistakes will be made. But if a person is essentially right, the mistakes he or she makes are not as serious in the long run as the mistakes management will make it if undertakes to tell those in authority exactly how they must do their jobs. Management that is destructively critical when mistakes are made kills initiative. And it's essential that we have many people with initiative if we are to continue to grow.

McKnight's statement was made near the midpoint of 3M's chronological history, and the subsequent years of substantial physical and financial growth have validated its wisdom. The company had already demonstrated vitality and a penchant for diversification in the years since its founding in 1902 as a corporation of five businessmen who set out to mine a mineral deposit near Two Harbors, Minnesota. Their idea was to produce material used in grinding-wheel abrasives, but that initial plan failed and the company moved to Duluth. It began the manufacture of sandpaper there in another uphill battle against established competition that controlled the supplies of the best raw materials.

On the verge of bankruptcy, the company found new investors and in 1910 moved to St. Paul, where its fortunes changed for the better. Corporate legend says its stock was once traded at two shares for a shot of bar whiskey, but Minnesota Mining and Manufacturing was able to pay its first dividend, 6 cents a share, in 1916. By the early 1920s the company proudly marketed its initial innovation, the world's first waterproof sandpaper, widely hailed as a safeguard against the dangerous sanding dust that plagued workers in a number of crafts. The invention of masking tape by a young 3M lab assistant in 1925 was the company's first deliberate venture into diversification, and that brought about a realization of the enormous potential outside the industrial markets that had been the center of its focus.

While much of the industrial world was mired in the Great Depression of the 1930s, 3M's creation of a whole family of pressure-sensitive tapes marketed under the Scotch brand label kept sales healthy and allowed it to expand its research operations. Out of this approach came new adhesives to replace upholstery tacks, soundproofing materials, and ceramic coating for roofing materials. World War II brought contracts for the production of materials for defense, and that research and development process inspired such innovations as Scotchlite reflective sheeting, magnetic sound-recording tape, filament tape, and new products to be used in the graphic arts.

It was abundantly clear to 3M's leaders before the war's end that innovative ideas in research could lead to new products that paid huge dividends. For confirmation, they had to look no further afield than the financial record of their wide assortment of Scotch brand adhesive tapes, which were the first in a number of accidental discoveries that produced lucrative profits. In an attempt to produce an adhesive masking tape to be used by auto body painters, a new lab man, Richard Drew, became overly conscientious in his devotion to economy and applied too little adhesive to the tape he prepared. In a test at an actual body shop, the tape fell off. The painter's criticism was brief and to the point: "Go back and tell your stingy Scotch bosses to put more stickum on this stuff!" Drew later succeeded in developing a better masking tape, and the auto painter's intended ethnic slur became the proud brand name for an entire new line of products.

The "Scotch tape" that truly became a household word and the beginning of a new industry was also created by Drew a short time later. His attempt to make a tape that would seal the slabs of insulation to the inside of refrigerated railway cars failed, but his experiment of applying a new rubber-based adhesive to moisture-proof cellophane led to an innocuous little product that has been a 3M best-seller from the Depression days until the present time. Scotch brand cellophane sealing tape was an immediate success with storekeepers, who, without the wide variety of packaging available to their modern counterparts, found it an invaluable aid for resealing packages to keep bread and other perishables fresh. When E. I. du Pont de Nemours & Company developed a heat-sealing process for cellophane packaging, company salesmen went in search of a new market. A salesman actually invented the tape dispenser that lessened the frustration of handling the sticky cellophane, and

3M sold the new all-in-one product to thrifty housewives intent on keeping food fresh and grocery bills low. Innovative uses for Scotch tape increased over the years, and its great success inspired the usual host of imitators, until the production of transparent sealing tape has grown into a billion-dollar industry.

The success of these and other early 3M innovations encouraged the company to expand and diversify. One of these technologies would lead to the discovery of Scotchgard fabric protector. In 1945, 3M acquired the rights to a process for producing fluorochemical compounds. At that time, despite its success, 3M was not a major research company, spending less than $2 million annually on that effort, considerably less than any one of the company's divisions spends today. Total sales in 1945 were $63 million derived from only four major product lines. CEO McKnight, R&D Vice President Richard P. Carlton, and Dr. Harry Stephens, director of 3M's Central Research Laboratories, all agreed that an investment of "patient money" in fluorochemical research might produce new products to add to the company's lines of coated abrasives, roofing granules, and adhesives and tapes.

It was a bold move at the time because very little was known about fluorochemistry outside the laboratories of Penn State University. 3M's decision to enter the field was a classic example of a farsighted management's opting to take advantage of academic research and transforming it into commercial products. The fluorochemical story began in 1936 when Penn State graduate student Joseph H. Simons dissolved acetic acid in liquid hydrogen fluoride and electrolyzed the solution, replacing the hydrogen atoms attached to carbon with fluorine. Simons continued to experiment and perfected it. About five years later he was given a patent on preparation of organic fluorine compounds by electrochemical fluorination. It was this technology that 3M purchased.

Dr. Lester C. Krogh, 3M's current vice president of Research and Development, who joined the company as a summer employee in the fluorochemical lab a few years later, feels the decision to enter that field was one of the best the company ever made. With its entry into an entirely new field of chemistry, 3M began to attract creative young research people who would achieve great success in a variety of areas in the years to come. The new effort also made it necessary to revamp the corporation's research department with the acquisition of new and better equipment. Included in that equipment was an entire new laboratory erected some ten years after the fluorochemical project started. Until that time all the research was done in an antiquated former liquor warehouse next to a railroad yard near downtown St. Paul.

There was great excitement in St. Paul about the potential of fluorochemistry, and that excitement made itself felt in other areas. When the company presented papers on its research at an American Chemical Society meeting in 1949, it attracted some potential customers and, more important, recruited more bright scientists eager to join the project. The handful of original researchers had expanded to thirty-five by 1950 and the roster reached one hundred by 1952. It was clear that 3M had embarked on a major research effort. What was far from clear was exactly what this effort was going to produce in the form of commercial products.

Originally, the best the early researchers could do was come up with low-boiling fluorocarbon gases and inert fluorochemical compounds. This success created excitement but nothing that could be marketed. With the addition of new people, the lab managed to produce some by-products, including reactive fluorine-containing materials that were highly unusual and among the most expensive organic chemicals known to man. But the all-important question for any profit-making concern remained unanswered: What, if anything, could be done with them?

After a huge investment and seven years of uneventful labor, management was forced to take a hard look at the situation. The fluorochemical project was by far the largest ever undertaken by the company, and it had produced virtually nothing that had made a positive impression on the well-known bottom line. Research for its own sake could be afforded only by academic or governmental institutions. Was it possible the project would never pay for itself? As head of the corporation, William L. McKnight was forced to examine that possibility. To his everlasting credit, he went about it in an innovative manner.

McKnight asked his vice president of Research and Development, Richard P. Carlton, to go directly to the source: to seek the educated views of those individuals most directly involved in the fluorochemical project. Carlton, like McKnight, still felt the project was capable of producing worthwhile results in time, but he stifled his own instincts in favor of putting some blunt questions to those who worked under him. Chief among those questions was, Should the company continue to pour a major portion of its resources into fluorochemical research?

Carlton called fifty of the scientists and managers working on the project into his office, one at a time, for personal interviews. His key question and the follow-up queries were equally to the point. He wanted nothing but honest opinions and rational arguments to support opinions. When his survey was concluded, he was able to report to

McKnight that forty-eight of the fifty people he had questioned all expressed solid confidence in the fluorochemical research and the rich results they expected it to produce. The project continued.

As the 3M researchers moved into their eighth year of work on fluorochemistry there were still no miracle solutions in sight. A business turndown in 1953 hit 3M hard, and the management opted to reduce the force working on the fluorochemical project to sixty-five. However, all of the thirty-five researchers taken off that project were placed in other jobs inside the company in less than two weeks. It was during this period, when the outlook for fluorochemistry seemed poor, that one of those marvelous accidents occurred—the kind of accident that can quickly turn the innovator's frustration into triumph.

Years of effort had enabled 3M's scientists to discover many unique properties in the new class of synthetic materials they had developed through fluorochemistry. Among the properties that fascinated them most was the insolubility of fluorocarbons in both oil and water. A number of teams were attempting to develop different products to take advantage of these properties. One such potential product had been the focus of their experimentation for a very long time. If they could develop a substance that could be applied to various fabrics, one that was resistant to both oil and water, it should have enormous commercial potential as a soil and stain resister. But despite continual experimentation, they had been unable to devise a polymer that would be colorless, odorless, and suitably durable when applied to a fabric.

Another 3M research synthesis group was attempting to create an oil-resistant fluoroelastomer as a rubberlike sealing material for use inside jet engines. A lab technician, working on the fluoroelastomer project, had the lucky accident that became a significant breakthrough for the "rival" group attempting to find a fabric protector. Patsy Sherman, like most of her colleagues, believed in dressing comfortably for her job in the stuffy old liquor warehouse that served as 3M's research lab and found tennis shoes to be a particularly practical addition to her working wardrobe. She was more than a little annoyed when, during a routine molecular weight determination, she accidentally dropped a sample she had been working with and it splashed profusely over one foot—clad in a recently purchased tennis shoe.

Worried about what the compound might do to the new shoe, she first attempted to wash it off with soap and water. When that failed, she tried to remove it with hydrocarbon solvents. They were equally ineffective. Although she was involved in different research, Ms. Sherman was aware that others were trying to discover an effective fabric protector, and it crossed her mind that the material she had dropped on her shoe had some of the properties the other group hoped to find. It was colorless and certainly resistant to water and hydrocarbon solvents, but she dismissed the thought and turned back to her own problems.

It wasn't until some time later that the physical evidence of her accident made her realize she had made a rather important discovery. The spilled material had not harmed the canvas fabric of her shoe and, when the new shoes began to show the grime and dinginess of the passage of time, the areas where the polymer sample had splashed were clearly delineated by their bright white color. The material had succeeded in repelling the natural dirt and grime. Fortunately, Sherman knew the exact composition of the spilled material and there were other samples available. She took her findings to Sam Smith, who was in charge of the laboratory.

THE INNOVATORS

DEVELOPING SCOTCHGARD FOR THE MARKETPLACE

Patsy Sherman's accidental discovery of a polymer material that resisted soiling proved to be the key breakthrough in the search for an effective fabric protector. On a higher level, it was the first practical indication that 3M's exciting and expensive program of fluorochemical research would eventually pay for itself. Still, it would be a long time before Sherman's discovery could be transformed into a viable commercial product. The process of developing the product, like the intricate research that had preceded it, was anything but accidental.

The material identified as the "sneaker culprit" actually did have many of the features the lab workers had been looking for, but it was hardly commercially adequate as it was. 3M's scientists embarked on an immediate program to develop a really efficient fluorocarbon polymer that would have better repellancy at a reasonable add-on level. With the shortcut provided by Sherman's accident, they could narrow the parameters of their experimentation and manage to demonstrate a thoroughly effective treatment that resisted stains within a relatively short period of time—compared, that is, to the years it had taken to reach that point.

But the process that was destined to inherit the famed Scotch brand name was still a long way from becoming a commercial reality. Fluorine-containing materials were incredibly costly, and used in its pure form the soil-resistant polymer was simply too expensive to be commercially viable. The company wanted to develop a product that could be applied to carpets and other fabrics at the mills where they were made. Obviously, the polymer would have to be *thinned* in some manner to make it less expensive and easier to apply while maintaining its effectiveness and durability.

76

Innovative action would be needed for this new trip into the unknown. No system that could do the job existed. 3M's researchers started from scratch to invent one, and many months of unsuccessful experimentation passed before they were able to succeed in that task. The successful emulsion polymerization process used water that contained a minor amount of an inert water-soluble organic solvent in which the fluorocarbon monomer became somewhat soluble.

Thus, in 1956, 3M was able to announce the arrival of Scotchgard Fabric Protector, a product that would eventually generate huge revenues for the innovative company that took its risky plunge into the mysterious world of fluorochemistry some eleven years earlier. The Scotchgard process was first sold directly to the textile and leather industries and became 3M's first volume product derived from fluorochemistry.

The Scotchgard label has now become almost as much of a household word as Scotch tape. In fact, 3M makes a constant effort in place to prevent the brand name from becoming a generic term. Its success led to 3M's establishing a new Protective Chemical Products Division that was split in 1986 into Protective Chemical Products and Industrial Chemical Products divisions. Scotchgard and another spin-off called Scotchban, a line of grease-resistant paper products, remain in the former division.

In its customary approach, 3M now markets a full line of Scotchgard brand products for the apparel and home furnishings markets. One of the more recent products to bear the Scotchgard name grew out of a question by a Sears executive who wondered whether the company could develop something to protect the surface of fine wood furniture. "We listened and challenged our lab to do it," National Sales Manager Jack Boyd says. "The basic technology already existed, since it's used in coating plastic lenses. After working for four years on the product, we successfully introduced Scotchgard brand wood protector in 1986."

From Patsy Sherman's accidental dropping of a test tube containing a fluorochemical mixture of unknown worth has come a new industry with chemical manufacturing plants in Alabama, Illinois, and Antwerp, Belgium, and main research laboratories in both St. Paul and Antwerp. And, after a slowdown in the mid-1980s, the Scotchgard Division is again meeting its goal of producing some 25 percent of its sales from new products.

THE ATTENTION TO INNOVATION AT 3M

3M had established a reputation as an *innovative* company prior to World War II, long before that adjective became the popular business buzzword it is today. Its bold venture into a new field of chemical research brought a number of key people into the fold, caused it to invest heavily in new research equipment, and, it can be argued, led to the building of the huge 3M center on the eastern edge of St. Paul.

Tom Reid, a co-inventor of Scotchgard Fabric Protector, set up a biochemical lab in Central Research some five years later. His ability to produce promising drugs led to the corporation's eventual acquisition of Riker Laboratories in 1969 and its commercial entry into the pharmaceutical business. Other fluorochemical research moved 3M into the agrichemical business to help farmers increase crop yields. The agricultural products project also produces herbicides and pesticides that control undesirable weed growths on everything from farm fields to golf courses.

The investment in fluorochemical research continues to pay big dividends for 3M, and the technology is interwoven throughout the many divisions that produce hundreds of successful products. 3M fluorochemical products are used to fight liquid fuel fires; fluoroelastomers that withstand extremes of heat and cold become seals and hoses in automotive engines; and polymer processing aids are incorporated into intricately molded rubber products as well as the huge molded devices used in oil drilling. The completely fluorinated organic compounds called inert liquids are utilized for semiconductor testing, vapor-phase soldering, and the cooling of aerospace radar tubes.

Still in the research and testing phase are programs examining the astounding potential of the high oxygen solubility of fluorochemical liquids, which may prove to be their most valuable property. It is possible that fluorochemical materials high in oxygen will someday enable man to explore the very depths of the ocean floor in safety. And the ability of fluorocarbons to carry oxygen has already prompted medical scientists, especially in Japan, to test them as emergency blood substitutes for humans.

Attention to innovation is at the heart of the "divide and grow" philosophy that has made 3M one of America's biggest and most diversified corporations. Although the company has made minor acquisitions from time to time, the overwhelming bulk of its growth has come from its own innovative projects. As in the case of fluorochemicals, the company expects the innovators in each of its technologies continually to spin off new products from the original line. Various technologies are linked to procure new commodities, and one of the company's stated financial objectives is to elicit a quarter of its annual sales from products introduced within the previous five years.

To ward off the threat of "bigness" that makes the management of true innovation impossible, 3M deliberately strives to miniaturize its working operations as it continues to grow. With forty-five major product lines, forty-plus product divisions, over fifty international companies, and more than 81,000 employees worldwide, this would seem to be an impossible task. But 3M has managed it by creating new entities when its divisions or subsidiaries grow too large. Only five of 3M's ninety-one American manufacturing plants employ more than 1,000 people and the median number for these installations is only 115. On the R&D side, the same approach is used, with the work divided into projects that rarely involve more than a dozen managers and professionals. This enables those with the special skills of the innovator-entrepreneur to approach their work as if they were heading their own company—but with the financial and technical backing of a major corporation.

The dangers inherent in bigness have caused the company to group its diverse interests along technology lines into four business sectors and a consumer products group. The latter group specializes in using the company's many technologies to produce products sold directly to the public, including such varied items as sandpaper and weather stripping for do-it-yourself chores, cleaning sponges for skin care, and fly-fishing lines for the angler. Scotchgard Fabric Protector is one of the 3M processes that is handled by the Industrial and Consumer Products Group, which also oversees and develops such products as tapes, adhesives, abrasives, and specialty chemicals. The Electronic and Information Technologies Group deals with products ranging from electrical connectors and videocassettes to office equipment and diskettes for computers. Graphic Technologies specializes in such products as photographic films, printing products, audiovisual goods, and advertising services; and the Life Sciences Group handles medical, surgical, orthopedic, dental, and pharmaceutical products as well as personal care, traffic safety, and personal safety products.

3M's management says the company's "divide and

grow" philosophy, its structure, and its traditions and values are all aimed at "encouraging our people to take an idea and run with it." The goal is to enable teams within a division to develop successful new products or businesses and break out the new enterprises into self-sustaining units, each with a large measure of responsibility for its own future. In this manner, projects can grow into departments and departments into divisions, while each of the semi-independent businesses continues to draw from the knowledge gained from research on the division, sector, and corporate levels. The success of this approach is evident in the fact that the company has identified and put to use about one hundred basic technologies, and 3M innovators are continuing to come up with fresh ways to combine them into new products.

A. F. Jacobson, chairman and CEO of 3M, admits that the company has not

> discovered a magic formula for producing entre-preneurs and innovators. In fact, there is a more or less continuous debate around our company about the proper way to stimulate new businesses and renew old ones. In my view, the development of entrepreneurs boils down to a principle that is fairly simple: Human beings are endowed with the urge to create, to bring into being something that has never existed before. That drive is stronger in some than in others. But, to some degree, it exists in just about everyone.
> It follows, then, that developing entrepreneurs simply means *respecting* that dimension of human nature and honoring it within the context of a profit-making enterprise. To me this means not simply encouraging innovation as an end in

itself. It means encouraging the conversion of innovation into profit-making businesses.

To further this aim, the 3M Corporation has developed specific management strategies that have become company policy. To encourage innovators, managers on all levels are instructed to offer them *challenge, responsibility, resources, sponsorship,* and *reward.* These five basic elements and all they entail constitute management's commitment to innovation.

The *challenge* part of the equation is quite clear in the company's published financial goals. Twenty-five percent of each division's sales are expected to be derived from products developed within the preceding five years. This is a tough challenge and one that not every division can meet year after year, but it encourages the managers to stay on top of new ideas that emerge from individual members of their own research and development teams.

Responsibility comes with each division's ability to operate much as a freestanding business in a clear attempt to eliminate many of the problems facing large organizations. The responsibility extends from the division general manager on down in 3M's standard procedure of utilizing teams of people from various disciplines who manage the product from the lab to the marketplace. Reasonable risk taking is encouraged and failure usually results in nothing worse than a chance to try again with another innovation. Close to 60 percent of the company's formal new product programs fail to meet expectations.

The third important element in 3M's commitment to innovation is the availability of *resources* for the creative thinker. It is firm company policy that products belong to the individual divisions, but technologies belong to *all.* Everyone in the entire company is encouraged to take advantage of them. It is this thinking that has inspired many of the "accidental discoveries" that have been developed into big money-makers. 3M uses this strategy to promote a great deal of cross-fertilization among the various technologies.

Sponsorship is the element in management strategy that receives an enormous amount of upper-level attention at 3M, because it is management's most direct method of proving that its commitment to innovation is more than just lip service. Innovators need both encouragement and support to transform good ideas into worthwhile products, and 3M expects its managers to provide them. Those in leadership positions are instructed to seek out those with innovative ideas, expedite their search for the needed resources, and personally shepherd a promising new project through the inevitable setbacks. They are encouraged to pay as much attention to patiently sponsoring new ideas and applications as they give to performing their standard functions as supervisor-critics for those who work under them.

Finally, those who labor to create innovations that eventually produce enormous revenues for their parent companies must be properly *rewarded.* This has always proved to be a special problem for companies that deal in scientific research. A fortunate few have extraordinary laboratory skills coupled with latent administrative abilities that allow them to climb the normal corporate ladder to better positions and monetary reward. But some brilliant technical innovators often prefer a lab workbench to a desk, choosing the great personal satisfaction of their research work over financial gain. 3M instituted its heralded "Dual Ladder System" a number of years ago to correct this situation. Everyone on the technical side of the organization has an opportunity to climb the regular ladder into positions as supervisor, manager, laboratory manager, and department director. But those with exceptional records of achievement in research and development who prefer to remain actively involved in their chosen field are given an opportunity to advance in parallel stages that allow them to receive monetary and other rewards on a par with their administrative colleagues.

The company in recent years has been experimenting with yet another ladder to reward. The new pathway, the Venture Career Track, is specially designed to appeal to those 3M people who have already given enthusiastic demonstration of their abilities as innovators and entrepreneurs. Instead of earning promotions on the other two ladders as a result of work completed, their rewards come as they move into even newer and more challenging areas—the heart of discovery. Concentrating on riskier emerging products, they have no need for second thoughts about continuing on the safer pathways to advancement.

It is clear that these and other management strategies have been specifically designed to foster a good climate for innovation at 3M. The company is known for its policy of permitting its technical people to spend up to 15 percent of their time working in the 3M labs on projects of their own choosing. Management admits that not everyone takes advantage of this option, but the ability to make full use of the company's technological resources to develop personal brainstorms has led to eventual company funding for innovations that have become commercial products.

One such product is 3M's *Post-it,* removable adhesive note pads that have produced a close to $100 million industry. This innovation came from an attempt of two researchers, Arthur Fry and Spencer Silver, to develop a superadhesive. What they came up with was a very weak adhesive that didn't interest the company. Fry, a church choir member, used a bit of it on a small piece of paper to mark his place in a hymn book and discovered that it could be easily removed and reapplied in other places. From that "religious" experience, Fry and Silver were able to convince management that they had created a new type of adhesive, and it quickly became a popular and profitable product.

An unofficial but time-honored offshoot of the policy that permits employees to use a percentage of company time to work on their own ideas is something that veteran 3Mers call *bootlegging*. Chairman Jacobson admits,

> Innovation is not a tidy process. There are always a few loose ends, things that drive the controllers a little wild. There are a lot of managers who like to keep things tidy with everything in its place and accounted for. So around 3M, what we call bootlegging has become a fine art. People working on new projects generally manage to beg, borrow, or scrounge up the resources they need. Somehow this activity does not always show up in the formal reporting system. As one 3Mer put it—there is such a thing as educating management beyond its need. We are comfortable with the situation and we encourage it.

3M, like other companies, spends a great deal of time and money keeping the lines of communication open among its many divisions. The corporation allocates about 6.5 percent of its annual revenues for research and development and considers communication an essential tool for making the most out of that investment.

Dr. Lester C. Krogh, 3M's vice president of Research and Development, says:

Because our forty-some divisions are fairly auton-
omous, there is always a tendency for technical
people to stay squirreled away in their own labs,
concentrating only on their own division's technology. To prevent this kind of insulation, we
maintain a massive and continuing effort to promote cross-communication among our technical
people. Through an organization called the Technical Forum, our people are in continuing dialogue
with each other. 3M's Technical Forum has more
than two dozen chapters and committees and may
stage more than 160 events in a given year, with
subjects ranging from "ion implantation in metals
and ceramics" to "new therapeutic approaches to
rheumatoid arthritis." The effect of all these seminars and colloquia is to keep hundreds of fresh
ideas bouncing back and forth among our laboratories all the time.

After a boom period in the second half of the 1970s when
its revenues doubled, 3M has felt increasing pressure from
foreign competition, particularly in its line of products
that store data, images, or sound magnetically. Despite
this and other economic factors that have affected most
U.S. companies, there has been a gradual upturn in 3M's
economic fortunes in recent years. Jacobson, a chemical
engineer who spent thirty-nine years with the company before assuming the chief executive's mantle in 1986, is convinced that breakthroughs in electronic and information
technologies will restore even that ailing sector to complete
health.

"I'm thoroughly convinced," he says, "that if we expect to stay competitive as individuals, as companies, as a

country, we *have* to have innovation. Genuine management commitment to innovation is one of the best guarantees we have for a bright future. It's not a cure-all, and I don't think it's just a fad or a buzzword. I've seen it work at 3M for a long time and we're doing our best to keep it working!"

5

Xerography:
The Revolution
in Office Work

"To set high goals, to have almost unattainable aspirations, to imbue people with the belief that they can be achieved— these are as important as the balance sheet, perhaps more so."
—JOSEPH C. WILSON

One of the maxims in any study of innovation is that small companies are invariably more capable of dealing with innovators and the products of their minds and hands than their larger counterparts. An unheralded physicist discovered the truth of that theory the hard way between 1938 and 1944 when twenty-one major American corporations expressed no interest in an invention he claimed would "make office work a little less tedious and a little more efficient." The invention seemed barely more than a demonstrable procedure, but a modest company in Rochester,

New York, saw enough potential in it to give it a try in 1948. Twelve years of innovative work and financial risk were required to translate the idea into a machine that would revolutionize office work—and transform the struggling firm into a respected corporate giant.

The original innovator, Chester F. Carlson, called his idea *electrophotography*, and only the Haloid Company of Rochester was bold enough to develop history's first "dry" office copier. That first machine's official designation was the Xerox 914 and its success was so great that the firm that produced it adopted its name and became the Xerox Corporation. A study of the heroic struggle that produced the "Xerox machine" illustrates both the risks and the rewards when innovation is allowed to flourish.

Machines of various types that copy or duplicate all manner of business documents are considered essential today, but their elementary forebears were a "hard sell." It took Chicago's A. B. Dick Company about five years to convince office managers at the turn of the century that Thomas A. Edison's original mimeograph machine was preferable to handwriting and typewriter carbons in making copies. The half-million or so mimeographs in general use by the start of World War II had little competition until office versions of the offset printing press began to appear. The offset device offered better-quality copies but also required the time-consuming and expensive preparation of "master copies" that made it unsatisfactory except for producing large numbers of reproductions. The expense and complexities of the Photostat and similar devices that operated on photographic principles and were available before the 1950s precluded their use in general office copying.

During the postwar boom a number of companies rushed to produce a machine that could make good copies. The innovative 3M Company introduced its Thermo-Fax in

1950, American Photocopy's Dial-A-Matic Autostat arrived two years later, and Eastman Kodak debuted its Verifax in 1953. All three found profitable markets despite some irritating shortcomings. None could produce copies except on special paper, Thermo-Fax copies became unsatisfactorily dark if exposed to too much heat, and the speed of Autostat and Verifax machines were offset by the time needed to dry their wet copies.

Although few were aware of it, Chester Carlson had come up with an idea that would solve most of these problems years earlier. Carlson's story of struggle and success has, in many ways, a ring of romance that seems lacking in accounts of many of today's innovators. He was the classic solitary wizard working in makeshift surroundings to create a device he saw in his mind's eye, a man never willing to admit that his dream might not be realized. Born in humble circumstances in Seattle in 1906, Carlson spent his formative years in southern California and worked his way through junior college and the California Institute of Technology, earning a degree in physics. He was hired as a researcher at Bell Laboratories, but his tenure there was brief. In the midst of the Depression, he was happy to land a job in the New York patent office of P. R. Mallory & Company, an Indiana electrical equipment manufacturer.

Carlson's work on that job started the flow of creative juices that resulted in his inventing electrophotography. As he explained it in a Xerox film of the 1960s, "In the course of my patent work I frequently had need for copies of patent specifications and drawings and there was really no convenient way of getting them at that time." Continual retypings of technical documents and expensive Photostats of patentable drawings were his only recourse. Believing he could find a better way, he enlarged his knowledge of physics and related subjects by spending his nonworking hours studying in the New York Public Library.

He deliberately bypassed possible photographic devices that experts in that discipline had failed to produce, but he was interested by the potential of photoconductivity and electrostatics. The electrical conductivity of a limited number of materials could be stimulated by exposing them to light. Carlson bought a supply of sulfur, the easiest material to obtain, and commenced work on a process that was beginning to take shape in his mind.

Still in his twenties, Carlson transformed his tiny bachelor apartment into a laboratory for his after-hours experimentation, using his kitchenette stove as a Bunsen burner and filling the halls of his building with the rotten-egg odor of sulfur smoke. His theory had promise, but his clumsy attempts to make melted sulfur adhere to a metal plate generated more neighbors' complaints than innovative progress. One of the complainers was a young woman who eventually became his wife. With a new bride sharing the polluted air of his lab-apartment, Carlson put his experiments on hold and refined the theory. It was only his limited lab skill that prevented his innovation from working, and he drew up a detailed patent application for electrophotography and filed it in 1937.

The marriage enabled Carlson to move his makeshift lab to a small apartment over a bar—in a building owned by his mother-in-law in Astoria, Queens—but he still couldn't assemble a working model of his invention. There was nothing to do but invest a portion of his earnings from Mallory on an assistant, and Carlson found a very capable one through a job-wanted ad in a physics journal. A refugee German physicist named Otto Kornei joined him to supply the laboratory expertise that Carlson lacked. Kornei's practical input lifted years of solitary drudgery from Carlson's shoulders, and, within three weeks of the older man's arrival, the technical problems had been overcome.

Eager to try his electrophotographic device, Carlson inked a terse message onto a glass slide that was one of the key components, along with a metal plate they had finally succeeded in coating with melted sulfur. Carlson rubbed the sulfur-coated plate vigorously with a handkerchief to stimulate an electrical charge on its surface, put the slide over it, and quickly exposed it by aiming the light of a floodlamp through the slide and its scrawled message. They dusted the plate with lycopodium powder and brought out the image of the message from the slide; when they placed a piece of waxed paper on the surface of the plate, the image from the plate was instantly transferred to the paper.

The content of the world's first dry, electrostatic copy was a model of simplicity, perhaps surpassing the other famous first messages in communications history for brevity. Carlson's said, "10-22-38 ASTORIA." It is now preserved in the Smithsonian Institution. Otto Kornei was not particularly impressed by their accomplishment and left soon afterward to take a better-paying permanent job in a large company. The innovator was alone again, his real travail just beginning.

Carlson began an enthusiastic search for a progressive company that would buy his invention. But his hopes dwindled when he received similar reactions from each one he approached, what he would later call "an enthusiastic lack of interest." Despite the disappointments, he continued working at his regular job and even took night-school law courses to learn more about protecting his invention. He received patents on several aspects of his process by 1944, including the design for a copying machine that he called an electrophotographic apparatus, but settled that year for an offer from a private research organization in Columbus, Ohio. The Battelle Memorial Institute agreed to exchange a $3,000 development investment for an eventual 60 percent of Carlson's future royalties. It was the best he could do in a six-year search.

Many scientists would say later that GE, IBM, RCA, and other companies had rejected his process because Carlson had created something unique, something with so little foundation in previous scientific work, that its value was not readily apparent. He had synthesized a number of poorly defined processes and made them work together, producing an effect that was remarkably unrelated to earlier scientific investigation. The Battelle people were as baffled as any but, through trial and error, ultimately added an improvement that made the process commercially viable. Carlson's sulfur-coated plate produced images that faded after a few copies had been reproduced. Battelle scientists wondered what would happen if they mixed a bit of a nonmetallic element called *selenium* with the sulfur as a coating for the reproductive plate. They tried it and the new blend worked better, even though they had no idea *why* it did. Adding selenium brought steadily improving results until they had completely eliminated the sulfur. The plate coated with selenium alone worked perfectly.

Carlson's association with Battelle cloaked him in a new mantle of respectability and he was invited to submit an article on his mystifying process to a magazine called *Radio News*. The piece created no significant interest until the research director of a struggling upstate New York firm came across a reference to it in a scientific journal he was reading a year after it was published. He was sufficiently interested to initiate a search for the back issue. What he read there interested him more—enough to set the proverbial ball in motion.

GAMBLE AND INNOVATION AT HALOID

After considerable thought, Dr. John H. Dessauer took Carlson's article to Joseph C. Wilson, namesake and grandson of one of the men who had founded the Haloid Company in 1906. Dessauer knew that Wilson, about to become the company's new president, wanted to find a way to lift the firm out of its postwar doldrums. The company manufactured photographic paper and related products with dwindling success against far stronger rivals, including its neighbor, Eastman Kodak, long the chief benefactor of Rochester's economy.

Dessauer convinced Wilson that Carlson's mystifying copying process was worthy of a close inspection, but, aware of Haloid's precarious financial situation, the future CEO was in no position to make rash decisions. A long period of thoughtful investigation ensued, followed by several trips to the Battelle labs in Columbus. The very first demonstration of Carlson's invention impressed Wilson and Dessauer, although neither understood the principles that made it work. More important, neither had any immediate ideas how the primitive device and strange process could be made into a money-making product. Wilson ordered low-key surveys to determine the commercial potential of a machine that could produce *dry* documents, and the results were generally encouraging. Dr. Dessauer believed his technical people could design a machine making optimum use of Carlson's process, and there was nothing on the current market that was even remotely as efficient.

Subsequent negotiations gave Wilson an estimate of the funds needed to obtain the rights to the process, and, when weighed in with the investment required to develop and produce the unique machine, the figures were frightening. But, after all the careful planning, Joe Wilson had become a believer. Fully aware that he was proposing a huge gamble that could end in failure, Wilson was able to convince his board of directors that the electrophotographic copier would put Haloid at the forefront of a new industry. He told the Battelle people he disliked Carlson's name for the process and, in consultation with a professor of classical languages at Ohio State, coined one that described the process in Greek: *xerography,* meaning "dry writing."

The agreement between Battelle and Haloid was reached in stages beginning in early 1947 and continuing through the fall of 1948. The first deal gave Haloid limited use of the Carlson patents, which was later expanded to give Haloid exclusive rights to the commercial application of the process. Haloid had two major obligations: to pay the heavy royalties that Carlson and Battelle would split and to share the work and development costs with Battelle. Exactly ten years after Carlson had reproduced the first xerographic message in his Astoria workroom—October 22, 1948—Haloid and Battelle sponsored a joint demonstration of his process at a convention of the Optical Society of America.

Haloid was forced to begin its work on the new copier without disrupting its basic income-producing work, and this in itself required innovative action. But the compactness and dedication of its technical group and general work force simplified the complex procedure, and close-knit teamwork enabled Haloid to introduce its first xerographic copier within two years of start-up. Prospective customers generally saw the Model A copier as a crude device that was too complicated for office personnel to operate. It failed as a copier, but the man who headed Haloid's sales department made it a modest success by finding an innovative application for it: as a quick maker of the master plates used in offset duplication.

The next Haloid copier was called the *Copyflo,* the

first automatic xerographic machine that could make copies on ordinary paper from either an original document or a microfilm image. It fared considerably better, and Haloid innovation was clearly producing better products. Company sales—from all product lines—had risen, reaching $21 million in 1955 and $25 million by 1957. But a large portion of those revenues went back into xerographic research and development; Haloid had not yet been able to produce the compact, easy-to-operate machines that Joe Wilson believed the market wanted. That would take more time and money, as would a planned big machine that seemed to have excellent potential. Wilson's closest adviser, attorney Sol M. Linowitz, had renegotiated the agreement with Carlson and Battelle in 1955 that gave Haloid full title to the patents. But the company was now obligated to pay all the costs of development.

Creative financial planning, enthusiastic salesmanship, and the goodwill of friends and employees provided the financial support that steered the Haloid Company ever closer to the goal that had lured it into its risky venture. Organization people mortgaged their homes to buy company stock and persuaded affluent friends to do likewise. The University of Rochester, in a gesture of community spirit, provided a lift by making a huge stock purchase. And Sol Linowitz negotiated a partnership with Britain's Rank Organization that created Rank Xerox Ltd. in 1956, opening up a European market for the company's planned super-copier.

Under Wilson's inspired leadership, Haloid was clearly beginning to make its mark. Dr. Dessauer's research group had been enlarged; Vice President John B. Hartnett had recruited and built strong sales, service, and marketing teams that kept the financial picture improving; and another VP, Harold S. Kuhns, continued to invent new money-raising schemes that permitted the xerographic development to proceed. By 1958, the success of its new product line was enough to inspire a change in the corporate name. After fifty-two years of business, Haloid became *Haloid-Xerox*.

The company seemed to be poised on the brink of the success that the leadership had anticipated when it obtained the rights to Carlson's patents ten years earlier. Innovative thinking rivaling that of Carlson himself had gone into the making of a prototype of a new machine that was capable of producing excellent dry copies, automatically and with dazzling speed. Workers from a variety of disciplines had practically invented the model as they went along—in shops that were dingy lofts above seed stores, garages, and even the basement of a local Masonic temple. Ingenuity compensated for lack of money, and some of the model's parts were salvaged from auto junkyards and other unlikely sources. An automobile tire pump inspired the method for lifting the paper off the new machine's rotating drum, the drum itself grew from a discarded drainage pipe, a local furrier trimmed the rabbit hair that would produce the static electricity, and the sooty men in the Masons' basement finally came up with a suitable toner. Next step: cost it out and go into production.

The estimated expense of the new copier surpassed anything the innovators had expected. The cost of refining the product, buying proper materials, setting up an assembly line, and hiring new people to man it was far more than the daring company could possibly afford. Expenditure was already out of balance with earnings, and finding new capital seemed impossible. He hated the idea, but, in appreciation of the board's patience and support, Wilson tried to find another way to get the job done. Discussions were begun with Bell & Howell about the possibilities of forming a joint venture: Haloid-Xerox would supply the technological expertise, and the big company would mass-produce the machines. Bell & Howell rejected the offer quickly, believing the process wouldn't work. IBM expressed mild interest when approached with a similar proposition but insisted on hiring a respected consulting firm to survey the market before making a decision.

The results of the IBM survey were extremely negative. The consulting firm claimed that the market found Wilson's machine too big and too expensive. Furthermore, there was no real need for it and Haloid would be lucky to sell a total of five thousand! That ended IBM's interest and stunned Joe Wilson, but he was not ready to surrender. He told his own board that Haloid-Xerox *had* to find a way to proceed. His findings were correct and IBM's were wrong. That survey hadn't really described the Xerox machine; its questions were more appropriate to the carbon-paper market. Haloid's market was those who needed copies that *didn't* cost 19 to 25 cents each. He won his case. Somehow, a way would be found.

BRINGING THE 914 TO MARKET

The third floor of another old loft building near a railroad siding on Orchard Street became the 914 assembly plant. Because the only access to it was a creaky set of stairs or two antiquated elevators, unsuitable for moving heavy parts and material, Haloid workers constructed a covered conveyor belt that successfully fed the parts and frames from a rear warehouse to the assembly staging area. There was no money for modern equipment; the first production models were built on wheeled wooden pallets, pushed along the line manually until each component was added.

On the occasion of the twentieth anniversary of the

Xerox 914, the man who had been the plant superintendent at the time fondly remembered the atmosphere. "Everybody—" John Klizas smiled, "managers, executives, foremen, hourly workers, engineers, everyone—had ideas and suggestions on how we could function better. And we listened and took action. It wasn't uncommon to work late into the evenings and Sundays, too. But people didn't seem to complain. Those were exciting times. We knew we had something good going for us, and we were all glad to be a part of it."

Work was often slowed by the difficulty of obtaining parts and, if there were no other options, the company chartered planes to collect the undelivered shipments. Other times, the company pulled cabinetmakers and other artisans off their regular jobs to make special fixtures that met the standards of those that were not delivered. The 914's engineers moved their drafting boards to the production floor so they could make instant design changes when trouble developed.

Ed Finein, a vice president, whose Business Products and Systems Group accounts for 90 percent of Xerox's revenue, joined Haloid as a truck loader in 1952—a college dropout with a pregnant wife. Finein, who completed his education in night school and rose through the ranks to become line manager for quality on the 914, remembers the firm's preoccupation with money problems.

> Merritt Chandler joined the company in the summer of 1959 to become program manager for the 914. I always say he was the one who had to teach the company how to spend money. I mean, we really thought *small* in those days. In 1958 I was the quality manager for the Copyflow machine that spat out twenty feet of paper a minute—on the

floor—and I needed something that would cut copy samples off it. I saw a rotary paper cutter that cost only $78, so I put in a requisition order. Someone calls two weeks later and starts berating me for wasting company funds! I finally calmed down and found out it was the *corporate vice president for finance*. He'd been guarding the money so closely that he was all upset about a $78 expense.

When we were doing the 914 and design or quality problems were keeping us from getting delivery of all the parts we needed—fifty to fifty-five major components—Merritt Chandler had come in to run the program. Merritt asked the head of procurement what he would need to make certain the parts got in, and the man answered that he'd need an engineer on location with each supplier for each part. Merritt said, "You'll have them tomorrow morning." And the next morning we had contract engineers assigned out to honcho those parts into the plant. Those were the kinds of changes necessary to drive a product of that magnitude into the marketplace.

Remarkable teamwork and planning triumphed over money woes and enabled the assemblers to produce five new 914s per day. The machine received its name because it could handle sheets of paper up to nine inches by fourteen inches in size, and it was big: weighing 648 pounds and at least the size of an office desk. It was not the compact model the market supposedly wanted, and many warned that heat and humidity would interfere with its paper handling operation in places like New Orleans. But the introduction of the Xerox 914 could not be delayed. Its reception would determine the future of Haloid-Xerox—one way or another.

Early embarrassing problems prevented the Xerox 914 from becoming a true overnight success. High humidity did sometimes cause the paper to stick to the photoreceptor drum; staples and other small office items clogged the mechanism when they fell into the machine; and a variety of small ills kept company repairmen on emergency standby to honor the product guarantee. Its bulk made it extremely difficult to move from place to place for personal demonstrations, so the company showcased the Xerox 914's simplicity of operation in television commercials that depicted a tiny girl running off copies for her businessman father. The TV spot convinced much of the business world that the 914 made copying "child's play"—that the push of a single button produced seven dry copies of a document on ordinary paper in just one minute.

By 1960, Haloid had spent $75 million on xerographic research and development—more than the company had spent on all its other products since opening for business, and *double* the income from its sensitized-paper line. There was clearly nothing on the market as good as the 914, but it faced strong competition from the established Thermo-Fax and Verifax desktop machines that sold for approximately $400 each. The bulky Xerox 914 had a price tag of $29,500. Competitive pricing figured to be a major problem, but Joe Wilson and his staff solved it with another innovative idea. You don't have to buy the product, they said; we'll *lease* it to you for $95 per month, and you can cancel the agreement on fifteen days' notice. We'll take care of any necessary repairs, give you the first two thousand copies free, and charge only five cents for each additional copy.

This idea not only permitted the 914 to demonstrate its effectiveness in a far wider marketplace but also proved to be a big money-maker for the hard-pressed company. The modest rental fees encouraged the start-ups of unexpected new businesses that encouraged individual users to copy documents at the corner candy store or stationery shop. It soon became a direct consumer tool in schools, libraries, post offices, and countless other venues in the United States and abroad. Once the initial glitches had been removed and the 914 had been given time to prove itself, it surpassed all expectations. The company committed itself to complete identification with its key product early in 1961, when it became the Xerox Corporation. Listed on the New York Stock Exchange that summer, Xerox reported year-end earnings in excess of $66 million.

Within five years of the 914's introduction, more than forty companies had entered the copier business, and industry revenue had tripled. It was estimated that the number of copies run off in the United States rose from 20 million in the mid-1950s to 14 *billion* by 1966. Xerox's annual revenues also climbed at a startling rate. Revenues were $33 million in 1959, but the success of the 914 nearly doubled them in 1960, elevated them to $166 million by 1963, and past the half-billion-dollar mark in 1966.

Xerox stock climbed to fifteenth place in market value that same year, and those who had purchased the nearly worthless paper during the lean years found that their investment had appreciated 180 times. Those who had risked their economic futures to back an innovation that nobody wanted had become a new breed—Xerox millionaires. The man who had started it all, Chester Carlson, was now a Rochester multimillionaire. Joseph C. Wilson and his associates continued to reinvest a large proportion of Xerox's revenues—$24 million in 1964 and more than $43 million in 1967—in research and engineering. Xerox produced several new winners even as the 914 was making its mark: the Xerox 813, the long-awaited tabletop model; the Xerox 2400, a copier/duplicator that could produce twenty-four hundred copies per hour; and the LDX, which allowed documents to be transmitted by telephone wires, microwave radio, or coaxial cable. The company also began marketing a similar device, the Telecopier, designed and manufactured by Magnavox.

The Xerox Corporation's burgeoning financial strength also enabled it to provide much-needed laboratories and factories for its growing work force on a thousand-acre tract purchased in nearby Webster, New York; another site near Rochester; and southern California. Its staff of scientists, engineers, and technical associates had been enlarged to several thousand and could work in comfort not even imagined just a few years earlier. In 1967, company executives and staff began moving into Xerox Square, its gleaming new international headquarters in downtown Rochester.

Innovative thinking and action from the boardroom to the factory floor had helped Xerox win its gamble and become *the* business success story of the 1960s. With success came expansion and diversification into advance information systems, education, and technology for outer space. Could a corporate giant maintain the enthusiasm, the quality, the innovative zeal that had propelled it to the heights? Before his retirement, Joseph C. Wilson confessed that even he had no sure answers to that question.

REINVENTING THE PROCESS AT XEROX

The Xerox Corporation was the unrivaled master of a growing new industry, producing new and better copier machines through the 1960s and into the 1970s, secure in the knowledge that it had exclusive rights to the Carlson patents. But a market as lucrative as the one Xerox had created was bound to attract other firms determined to get a slice of the financial pie.

The first worrisome challenge came in the early 1970s from IBM, which began concentrating on the midrange market, which was and is the most profitable in the business. There was no way to produce a true xerographic machine without using Chet Carlson's basic approach, and Xerox sued IBM for patent infringement. But Xerox was hit in 1972 with a Federal Trade Commission suit for violating antitrust laws, and that brought another the following year, filed by the SCM Corporation. Xerox won against IBM and SCM but lost the federal antitrust suit in 1975. It could no longer refuse to license its xerographic know-how.

Xerox began to feel the effect of that ruling increasingly during the late 1970s. Japanese firms had begun importing cheap, low-quality copiers aimed at the lower end of the U.S. market in the early 1970s. It had little effect because Xerox generally ignored that market. But the situation changed rapidly in the mid-1970s when a few Japanese companies introduced new copiers that elevated xerographic technology to a new level, certainly in terms of cost and simplification. By 1979, Xerox was engaged in a major struggle with Japanese competitors. At least six major suppliers and twice as many minor manufacturers were capturing huge portions of the marketplace. Xerox's share of the market had fallen from 95 percent in 1970 to barely 50 percent in 1979.

The leader was steadily being toppled and, rather than throw up their hands in dismay, the Xerox people set out to pinpoint the problems and correct them. The first step in what would become a reinvention in the process of innovation—competition benchmarking—was taken by the company's manufacturing group. In a strictly financial investigation to learn how the Japanese could produce high-quality machines and sell them at low cost, a delegation of economic experts from the manufacturing group traveled to Japan to examine the competition's overhead rates, commercial costs, lead times, and similar information. This prompted an in-depth study that compared Xerox, point for point, with Japan's best.

The findings revealed that Xerox spent about twice as much as the Japanese to make a comparable copier. The quality of Xerox products trailed the Japanese by a factor of 10. Xerox took twice as long to get a product to the marketplace, although it employed twice as many engineers. While Xerox had been improving its overall performance by 8 to 10 percent per year, the Japanese were demonstrating annual improvement at a rate of 12 to 18 percent. And in terms of the most important benchmark of all—customer satisfaction—Xerox was astounded to learn it had *no* product at the top in customer satisfaction in any market segment.

Drastic measures were called for and, after careful planning, in 1982 the company was restructured into business units with a strong chief engineer at the head of each. The experienced chief engineers replaced the program managers of the old system who had tried to coordinate product development through "committee meetings" that soon degenerated into parochialism, interdepartmental rivalries, and general loss of goals and direction. In the new setup, chief engineers had direct-line responsibility for designing, manufacturing, and delivering products that would be acceptable to the customer. The chief engineer was allowed to become an entrepreneur with the authority to make decisions and drive his product into the marketplace, a power no single individual had ever been given in the past.

The company intended to achieve what it called *business effectiveness*—with sweeping improvements in customer orientation, competitive benchmarking, employee involvement, and total quality. Restructuring had helped, but more long-haul improvements were needed. Even if the company succeeded in dropping its manufacturing costs to the Japanese level, marketing expense would still make Xerox products twice as costly as the Japanese. Was it a no-win proposition? Progress was painfully slow, causing David T. Kearns, Xerox chairman and CEO, to swallow his pride and consider yet another option. Was it possible that American business needed to learn from the Japanese, long labeled the "copycats" of the industrial world? It was worth a look.

Kearns took the bit in his own teeth and traveled to Japan to study operations at Fuji Xerox, the company's affiliate that had just won Japan's prestigious Deming Award for its success in instituting a total quality process within its own organization. Kearns was immensely impressed with what he found and returned to Rochester to formulate plans for starting a similar but more formidable process at Xerox. He planned an awesome undertaking, one that would challenge American workers to relinquish familiar procedures of the past in favor of new methods that might allow them to compete successfully with ambitious foreign competitors.

The new program, Leadership Through Quality, was made company policy in 1983. Designed to develop and implement a total quality process within Xerox, it necessitated one of the biggest retraining programs in U.S. business history. Every individual in the organization would receive training that explained the meaning of world-class performance in his or her occupation and would learn methods that would make that kind of performance possible. It was explained to all that Xerox had to elevate its performance and products to world-class levels or face failure. The goal of the companywide program—mandatory for all occupational levels, all disciplines, all *people* within Xerox—was nothing less than creation of a completely new corporate culture that would make the company competitive with any other in the world.

The training program began at the very top and started flowing downward through the organization in what the company calls a "waterfall" effect. Kearns and his staff were the first to be instructed, they trained their respective staffs, those people passed the information on to theirs, and on it went, down the line. Each employee received at least a week's training—in practical workshops and seminars— and each family-group training session was conducted by a directly involved superior. No outside consultants were used because management believed that Xerox people were more knowledgeable about internal problems. By mid-1988, all 99,032 Xerox employees had received at least a week of quality training.

Ed Finein, vice president and chief engineer of product delivery processes, has been directly involved with Xerox's efforts to reinvent the innovative process that had made it successful.

One of the cornerstones of Leadership Through Quality is the concept of utilizing the resources— the brainpower, the energy, the knowledge—of all the people in the organization. Employee involvement certainly leads to group innovation, and it strengthens the climate for individual innovation. We have a number of approaches for fostering innovation within the company. We encourage people with good new ideas to go to an ombudsman, as it were, who controls special funds we have set aside that allow them to get financial backing to follow through on that idea or that approach. We

also have so-called Innovators' Fairs in several locations where groups and individuals who have
achieved unique things can come and display their
work and receive recognition. And we have a few
"skunkwork" kinds of operations where we've
given people support and told them to go out and
do whatever they can to develop new market segments or products or whatever.

The company rewards innovative thinking in many ways and
can document employee innovations that have saved the
company millions of dollars. Does the Leadership Through
Quality program work? Since its introduction in 1983, Xerox
has cut manufacturing costs by more than 50 percent and
demonstrated a tenfold improvement in the quality of its
products. Surveys show Xerox at or near the top in customer
satisfaction in almost every market segment.

Xerox is not ready to claim victory over its Japanese
rivals, but the Leadership Through Quality approach has
produced a steadily improving situation. Annual revenues
increased by $5 billion in the 1983 to 1987 period, and mar-
keting statistics show Xerox recapturing segments of the
market lost during the late 1970s and early 1980s. Xerox
unveiled its new 50 Series of copiers in 1988, honoring the
fiftieth anniversary of Carlson's invention of xerography.
One model sets a new speed record of 135 copies per minute.
Another, a personal copier, bluntly challenges all competition, domestic and foreign, to equal it.

"This structure," Ed Finein says, "has brought
back—to a very large degree—some of the innovative spirit
that I saw here in the late fifties and early sixties."

6

The Coming of the Laser: From Operating Room to Supermarket

"When something radically new comes along, it takes a while to find a proper use for it. People forget that it was about thirty years between the Wright brothers' first flight and the start of commercial aviation."

—DR. THEODORE H. MAIMAN

Innovations frequently spring from the collective efforts of many, often within a single organization, but the development of the laser provides an opportunity to study a different aspect of the innovative process. *Different* is the operative word because the "laser" is not a single entity but a wide variety of devices created in a number of laboratories, academic and industrial. It is an instrument of such unique power that finding commercial uses for it remains an ongoing challenge, and even its very beginning was, for a long time, obscured in controversy.

Its name, *laser,* comes from the acronym of the function the instrument performs: light amplification by stimulated emission of radiation. And, once again, science must credit Albert Einstein for developing the theory that others were able to apply later. Believing that spontaneous emission occurs when an atom in an excited state eventually decays of its own accord and emits a photon of energy as it does, Einstein theorized that this condition could be stimulated by hitting the excited atom with a photon of equal energy as the one being emitted by natural means. This, he calculated, would enable two photons of like energy to exit together—traveling in the same direction, exactly in phase with each other.

Although the theory was brilliant, it would be many years before scientists could think of applying it. The speed at which the process occurred, sometimes less than one-hundred-millionth of a second, was beyond human control. A part of the theory was finally demonstrated in 1954 when three Americans invented what they called the maser: microwave amplification by stimulated emission of radiation. Charles H. Townes, James P. Gordon, and H. J. Zeiger, using Einstein's principles, succeeded in amplifying *microwaves*—electromagnetic waves produced by transitions between the energy levels of molecules in the same manner that light is produced by atoms.

Townes had first begun exploring microwave spectroscopy as a researcher at Bell Laboratories in New Jersey and retained a consultancy role with Bell after moving on to a professorship at Columbia University. The maser that he, Gordon, and Zeiger created grew out of graduate student experimentation at Columbia, and its introduction stimulated real interest in producing *light* in the manner that Einstein had charted many years earlier. Because microwaves and light waves have much in common, the operation of the maser provided vital insight into ways that atoms might be manipulated to produce light. Scientists in laboratories around the world began exploring the possibilities, leading to much of the obfuscation of the beginnings of the complex laser story.

The original history of the period credited Townes himself as co-author of the first great breakthrough. In 1958 he and Bell Labs physicist Arthur L. Schawlow, who had studied under Townes as a postdoctoral student at Columbia, published a paper in which they outlined their ideas and methods for amplifying light waves by stimulated emission. It was the first such paper published and was widely hailed in the scientific community. They were awarded patents for the invention, and Townes would later share a Nobel Prize in Physics with two Russians for work leading to the development of both masers and lasers.

Even as scientists around the world began following the principles laid out by the Schawlow-Townes paper, less publicized events began to complicate the orderly chronology of the laser's early development. In November *1957*—months before Townes and Schawlow revealed their findings—one of Townes's Columbia graduate students reached similar conclusions on his own. Gordon Gould, a brilliant thirty-seven-year-old with a bachelor of science degree from Union College and a master's in physics from Yale, was convinced he had conceived something of immense importance and knew that he had to protect it. His plan for producing a beam of coherent light was carefully formulated in his notebook, and he ran to an off-campus candy store to have his description of the concept notarized. He hoped this would be sufficient to give him proprietorship of the process he called "light amplification by stimulated emission of radiation."

Gould has since admitted that he should have immediately filed an application for a patent, but, at the time, he thought it was necessary to have a working model of an invention before it could be patented and made no attempt to file until 1959. More frustration followed when he learned that his Columbia professor and Schawlow had already been awarded laser patents that had become the property of Bell Labs. Gould believes his first application "had ten different inventions in it" and, although most of them differed from that of Townes and Schawlow, the U.S. Patent Office policy allows the issuance of only one patent for each invention. His only option was to contest the issue in what the patent office calls "an interference," a war of attrition that would drag on for many years.

The process that Schawlow and Townes had patented envisioned the use of potassium vapor, enclosed in a glass tube, as the active material that would make laser action possible. The theory seemed sound because the greatest percentage of the volume of a gas is actually empty space, making it far easier to predict how it will behave when an attempt is made to stimulate the emission of its radiation. Each molecule can be considered individually without undue worry about the influence of its neighbors. Potassium seemed a particularly good choice because its supposed molecules are actually single atoms, making the calculations of the gas' behavior even less complicated. But when Schawlow and Townes attempted to put their theory into practice, they failed. They couldn't energize the element enough to produce emission.

Many other scientists also experimented with a variety of monatomic gases in attempts to build a physical model of the device that Schawlow and Townes had described, but the race to produce the first working laser was won by a surprise entrant who was far removed from the early excitement—a man with a new and different approach. Dr. Theodore H. Maiman, a physicist at Hughes Aircraft Research Laboratories in Malibu, California, displayed the zeal of the true innovator by battling financial and scientific odds to demonstrate the first light ever amplified by the stimulated emission of radiation. His persistence paid off in the spring of 1960, when, for approximately a thousandth of a second, he produced a beam of light that outshone the sun.

When interviewed for this book, Dr. Maiman reluctantly conceded that his employers at Hughes Laboratories were "not very cooperative" during his determined search for the laser. Not wanting to denigrate a former employer, Maiman says Hughes was understandably reluctant to plunge into laser research because it was already doing well with microwave technology and had numerous government contracts for radar devices and other military electronics. General research funds came from corporate profits, and the management was wary about investing in anything so revolutionary and unknown. A study of the situation reveals that the conflict between inspired innovator and conservative leadership can still result in a profitable expansion of company horizons.

Maiman had been working on masers at Hughes under a government contract since 1957, and he began to formulate his own ideas about the best ways of producing a laser during that period. He describes his early interest in an optical device as only "background thinking," but he had clearly become involved enough to disagree with the theories expounded by Schawlow and Townes when their paper was published in 1958. He remembers thinking at the time that their own calculations pretty well demonstrated that what they were proposing wouldn't work. Rather than influencing him to change directions, the Schawlow-Townes report reinforced his belief that his simpler approach was more likely to succeed.

Although he had done some qualitative thinking about the possibility of creating a laser, Maiman did no practical work on the project until August 1959 after he had delivered a miniaturized maser he had produced to the Army Signal Corps. Less than nine months later, he would unveil the first laser. Maiman feels that incredibly brief interval between conception and delivery is a bit misleading. His academic and professional background made him especially well equipped to do the job. It was also instrumental in gaining the company's permission to begin it.

My background was probably one of the key reasons I was the first to be successful. It required a mixture of backgrounds, and most of the researchers up to that point didn't have the combination. It required some knowledge of electronics as well as physics and optics. I happened to have some background in electronics—my master's was in electrical engineering—and I had worked in electronics since I was a boy. My Ph.D. was in physics at Stanford, and that did entail the use of optical techniques. I had some background in vacuum systems—in fact, my Ph.D. thesis was on a gas—and some of my first work at Hughes involved vapors. I had worked in solids with the ruby maser, so I was familiar with the optical properties of gases and solids and vapors—and the electronics. And, of course, I had gotten a lot of solid practical laboratory experience in addition to the analytical training from my Ph.D. So, I guess you could say I had the proper tools.

Illustrative of Dr. Maiman's innovative courage—and a contributor to his company's reluctance to give him unbridled support—was his determination to *go against the grain*! While better-known physicists in labs around the world were attempting to induce laser action from monatomic gases and vapors, Maiman was convinced that light amplification could be achieved by exciting the atoms of a *crystal*. His work with ruby crystals in the maser made it his first choice as the active material in an optical device, but he became sidetracked when he came across some published measurements that indicated the ruby would be totally impractical for optical purposes. That led him into some exploration of other solids that also proved unsatisfactory.

He returned to the ruby because its properties were all very promising, except for one negative that had been publicly recorded. He decided to make his own detailed measurements of the ruby, not so much to prove its suitability as to understand the reason for the "bottleneck." This, he hoped, might help him find a more effective crystal. Ruby is a fluorescent material that casts a deep red glow under a beam of ultraviolet light or, as Maiman discovered, even a beam of blue or green light. The scientific measurement of the amount of red that is released when a light is aimed at the crystal is called *fluorescent efficiency*. The problem with the ruby, according to the published report, was its minuscule fluorescent efficiency—measured at about 1 percent. Attempting to discover why the ruby's fluorescent efficiency was so low, Maiman made many measurements of his own and was shocked to learn that his findings gave the ruby an efficiency of *70* percent.

As Maiman says, there are a number of ironies in the story of the laser's development, and one of them concerns the erroneous report that had convinced physicists everywhere that the ruby was totally unsuitable as the active material in a laser. Maiman was particularly surprised to discover the error because he knew the individual who had written it and had considered him a protégé during his time at Stanford. His discovery of the huge mistake reinforced his original thinking and made him even more eager to proceed. Hughes had backed him with a modicum of general research funds but advised him to apply for a federal contract if he hoped to continue. Maiman complied but received no answer.

In timing that coincided with his reappraisal of the ruby's fluorescent efficiency, news came from an important meeting of physicists in upstate New York, which brought new problems for Ted Maiman. Arthur Schawlow reported that he and his colleagues at Bell Labs had also been examining the possibilities of the ruby crystal and had found it totally unworkable. Schawlow and Townes were considered *the* experts on the laser, and the former's damning of the ruby convinced management that their group leader in the atomic physics department was flogging a dead horse. The company asked him to abandon the project.

But Maiman was vehement in his own defense and insisted that it was Schawlow who was in error. It was his opinion that both the original Schawlow-Townes laser paper and Schawlow's new statement demonstrated only that they lacked "good solid lab experience." They had been wrong in both directions, he insisted: wrong about the vapor because they claimed it would work and wrong about the ruby's not working because Schawlow failed to see that the problem with it was solvable. Maiman agreed that it would be tough to produce laser action from a ruby, but he *knew* it could be done, with sufficient excitation.

The strength of his argument, coupled with his proven expertise in the development of the maser, won the day. Maiman viewed his success as a "reprieve" rather than an enthusiastic go-ahead, aware that his superiors knew that some world-renowned scientists claimed that a laser would never be made. But the company bought a $1,500 instrument he needed and allowed him to continue.

As 1959 drew to a close, Maiman put together a model of the instrument he hoped to create. His plan was to shine a light on the ruby to make it fluoresce at a very intense level until it became a laser. He calculated the parameters of the light source needed to excite the ruby and confirmed that the total power of the source would be secondary to its *intensity*: the power per unit area in a given color range. An intense light source would make it unnecessary to have a large ruby, and his systematic search for a suitable lamp raised some new concerns. He almost settled on a mercury arc lamp but worried that it might be only marginally effective. What if it didn't work? Would the naysayers smile smugly and scrap the entire project? He eventually selected a seemingly prosaic item: a photographic strobe light. Research on flash lamps indicated that only three seemed capable of producing the intensity he needed. The smallest of the trio, General Electric's FT-506 xenon flash lamp, should do the job.

Linde, a division of Union Carbide, grew man-made ruby crystals for use in synthetic jewelry, and Maiman intended to use one of them for his laser. The properties of these gems were identical to those of natural rubies and were better for his purpose than those taken from the earth. Union Carbide shipped the crystals to Hughes as raw material and Maiman, conscious of his limited budget, asked the Hughes tool shop to cut and polish them. Unfortunately, Hughes was not equipped to provide the mirrorlike finish that would be most desirable, but he proceeded to design his experimental device with what he had.

By spring 1960, Maiman was conducting optical-microwave experiments that revealed a significant decrease in the microwave absorption, indicating that an appreciable fraction of the ground state was being depleted. When Maiman was nearly ready to begin his laser experiments, Hughes Labs became involved in a mammoth moving project—vacating the old premises in Culver City and establishing new headquarters at Malibu. It was something over two weeks before Maiman and his colleagues were able to get organized and restart the now promising experimentation.

The device that Maiman had spent months thinking about and making in the hope of producing history's first laser beam was deceptively small and simple-looking in design. In fact, Maiman admits that many early pictures of the ruby laser were, in his words, "totally phony." A Hughes PR photographer considered the actual device insufficiently impressive and insisted on posing him with related odds and ends from his lab. The real laser consisted of his pink ruby cylinder from Union Carbide, 2 centimeters long and 1 centimeter in diameter, with both ends ground and polished flat and parallel. The ends of the crystal were coated with evaporated silver to make them act as mirrors, and one end had a 1-millimeter hole in the coating to couple out the radiation. The ruby was mounted on the axis of the lamp with its spiral flashtube surrounding it to produce sufficient energy to excite the atoms inside.

Dr. Maiman's date with destiny was May 16, 1960, when he achieved the first laser action from his ruby crystal. He had feared that the inadequate polishing of the crystal might interfere with the production of a visible laser beam and had ordered three more rubies, properly ground and finished, from Union Carbide. Maiman used a crystal polished by the Hughes tool shop. His laboratory instruments all verified the fact that his device was delivering an incredibly intense and coherent laser beam through the partially silvered end of the crystal.

FINDING A NICHE IN THE MARKETPLACE

Aware that he had created an unusual tool that would surely have many kinds of uses in the future, Maiman wanted to follow standard scientific procedure and publish a paper detailing his accomplishment. The company was eager to gain the prestige from such a momentous scientific achievement, and George F. Smith, who was later to become director of Hughes Research Laboratories, describes what followed as "a comedy of errors." Maiman fired off an article titled "Optical Laser Action in Ruby" to the respected scientific journal *Physical Review Letters*. It was rejected because the journal ruled that it had run a surfeit of "maser" stories. Maiman's paper, retitled "Stimulated Optical Radiation in Ruby Masers," was eventually published by a British journal, *Nature,* three months after he had demonstrated the laser.

Unwilling to wait longer because of fear that others were close to producing a laser, Hughes scheduled a full-scale press conference at New York's Delmonico Hotel for July 7, 1960. That, in Maiman's opinion, was the company's worst blunder to that point. It placed him in an unfavorable position with the science journals, who had a policy of demanding first publication rights; more important, it immediately cost Hughes the worldwide rights to the ruby laser. Hughes had not yet applied for a patent, knowing it had a full year to do so after first publication in the United States. But that rule did not apply in most other countries. Maiman claims the company did not file for a patent until he announced his intention of leaving, in April 1961.

The press conference produced more results that were unsettling to Maiman. There was an exceptional turnout of respected science reporters representing newspapers and magazines around the world. Maiman gave a detailed speech on his innovation, making sure he referred to it as a laser as well as an optical maser, and displayed a model that could be inspected at close range. But the tone of the conference changed during the informal question-and-answer period.

> Several things happened there. The writer from *Time* magazine got up, slammed his press kit on the table, and said something like, "What are you guys trying to pull here?" He apparently hadn't believed anything we said. *The New York Times* writer tried to explain it all to him and that made him even more angry. He got all red in the face and everything, thinking the older gentleman was talking down to him. *Time* magazine became the only major publication that didn't run the story, although it made all the other magazines and the front pages of just about every newspaper in the country.
>
> After I had stepped down from the podium, several reporters gathered around to ask some more questions. One man, from the *Chicago Tribune,* asked if it could be a weapon. Well, I pointed out that I had mentioned many possible uses—in communications and as a scientific tool for industrial, chemical, and medical purposes—but I hadn't thought about it as a weapon. He kept after me— it was my first encounter with the media—and I

finally said I guessed it *could* be used as a weapon, maybe twenty years in the future. He said, "That's all I wanted to know." The next day a Los Angeles paper had *red* headlines on the front page—a couple of inches high—saying, "LA MAN DISCOVERS SCIENCE FICTION DEATH RAY." Just about all the headlines across the country were some variation of that theme. The only paper to cover it responsibly was *The New York Times,* which also put it on the front page but didn't say anything about a death ray.

The optically polished rubies from Linde arrived at Hughes a few days after the press conference. Maiman popped them into his device and got a strongly visible laser beam. Hughes had won the race, but work on other lasers was continuing in many laboratories. Toward the end of 1960, a Bell Labs team headed by Ali Javan made the first gas laser. Javan's laser was strikingly different from Maiman's, using a mixture of 90 percent helium and 10 percent neon gas as the active material and a small radio transmitter to pump the helium atoms until they reached an excited state. Javan's innovation spurred the creation of lasers in several laboratories using a variety of other noble gases in 1961 and 1962. The problem with most of these devices was their inability to produce *visible* laser beams, meaning they emitted infrared beams, which are less powerful.

A significant advance in laser development came from innovative research by American W. E. Bell, who used a different gas, mercury, to produce a visible laser beam. Bell was able to achieve the desired results because he had altered the composition of the mercury gas until its normal atoms were left with a small net positive charge, transforming them into *ions*. His work led to the development of other ion lasers that were effective as sources for powerful red and green laser beams, although they proved to have short life spans. The earliest of these lasers were pumped by electron beams that also ionized the atoms in the active material. The enormous amounts of heat created during the process limited the power and led to deterioration, as did the constant pounding of the cathode and walls of the tube by the positive-charged ions. Better pumping methods were found later.

Many different types of lasers were being made, including a significant advance from Bell Laboratories in 1965. C. K. N. Patel demonstrated a molecular gas laser there that was more powerful and more efficient than previous gas lasers. Patel devised a new system called the *flowing-gas structure* in which gas with heavy molecules was kept flowing between two mirrors, where it was hit with a flow of already-excited nitrogen gas. The collision of the two gases allowed the energy of the nitrogen atoms to excite the heavy molecules of the active gas chosen. Patel was able to demonstrate laser activity with nitrous oxide, carbon disulfide, carbon monoxide, and carbon dioxide. The carbon dioxide laser proved to be the most efficient and, within two years, such organizations as North American Aviation, Inc., the Raytheon Company, and the Massachusetts Institute of Technology were demonstrating powerful results with carbon dioxide lasers that showed promise of becoming the workhorses for industrial, military, and research purposes.

Hoping to find a laser that would be smaller and more mobile and would not need a heavy and cumbersome power supply, some groups in the early 1960s examined the potential of semiconductors. Scientists had been fascinated for years by the bright red light emitted from a gallium phosphide semiconductor when a current was passed through it, and work to transform that light into a laser beam began shortly after Maiman produced his ruby laser. By 1962, independent research by teams at three different organizations—General Electric, IBM, and MIT's Lincoln Laboratory—produced semiconductor lasers. Steady improvements in active materials and pumping methods gave semiconductor lasers great promise for a variety of uses. Their output power could be controlled easily by the supply voltage, making them suitable for carrying sound and pictures—a portent of things to come.

Laser development was proceeding rapidly in the years following Maiman's breakthrough, but commercial application of laser technology was lagging far behind. The general public, as well as many in industry and business, couldn't shake the idea that the laser was anything more than a potent instrument of destruction. Dr. Maiman remembers a chance encounter that illustrates the point.

> I was introduced to Bette Davis at a party and the first thing she said was, "Gee, how does it feel to be responsible for that destructive machine?" That really took me by surprise, and I'm not sure how I responded, other than trying to explain that she had some misinformation. But later, when she was leaving, she made a point of coming over to me and said that she felt she had been unfair. She let me off the hook on the basis that it was up to society—when scientists come up with something—to handle it responsibly. There was so much of that in the early days that I became pretty defensive about it, but I still don't know of anyone who's been killed by a laser, even accidentally.

The military was clearly interested in the laser's destructive potential. Maiman's earlier application for a federal contract was eventually approved, and Department of Defense funds undeniably helped speed laser development. A 1960s plan to use laser beams against enemy missiles was dropped as impractical, only to be resurrected in the Strategic Defense Initiative of the Reagan administration. No known death ray exists, but the military now uses lasers for many purposes and federal funding continues to provide a vital fringe benefit—better lasers for humanitarian uses.

Initial alarm about the laser is understandable in hindsight. Its beam of coherent light had the power to burn a hole in a wall or a plate of steel if focused narrowly, but scientists saw it only as a remarkable *tool*. Its raw power was less important than the qualities that made the power possible. To the naked eye, the sun appears to give off light in the yellow-red-orange portion of the spectrum. In reality, both the sun and the ordinary white light bulb shoot off light in a wide range of colors or wavelengths that radiate in all directions from the source. The laser produces light waves of all one color and wavelength, and these waves move in unison in the same direction. Its uniformity and narrowness of focus make it special.

Securing the laser's niche in the commercial marketplace was complicated by the shortage of complementary technology. Light wave telecommunication was an old idea that now seemed possible if the laser were used as the light source, but no medium for carrying the light arrived until ten years after Maiman made his ruby laser. Most scientific innovations take time to move from the workbench to the marketplace; because of its absolute uniqueness, the laser would take more time than most.

One of the first practical demonstrations of laser use was given by a physician at Children's Hospital in Cincinnati. Dr. Leon Goldman applied it in the treatment of melanoma, a form of skin cancer, in 1961 and also proved later that it could successfully burn away port wine stains, tattoos, and other discolorations on human skin. Ophthalmologists were particularly eager to find out whether the ruby laser could replace the unsatisfactory xenon lamp as a source of light energy for "spot welding" detached retinas. Surgeons saw a host of possible uses for the laser in medicine, but these applications could not be put into practice until devices that would give them meticulous control of its power were made.

Laser production seemed destined to become a lucrative new industry, but the nature of the product dictated the need for innovative planning. The key to gleaning the anticipated financial harvest was the tailoring of a wide variety of lasers to perform specific functions. The most common use for lasers in the 1960s was in areas unfamiliar to the lay public—as superb tools that performed countless laboratory jobs previously thought impossible. Meanwhile, the public was dazzled by the ability of the laser to transmit three-dimensional holograms and watched in awe as two laser beams hit the surface of the moon in 1968—a year before Armstrong, Aldrin, and Collins made the trip. But significant progress in laser technology wouldn't begin until the 1970s.

INNOVATIVE APPLICATIONS OF LASER TECHNOLOGY

Few events electrified the fledgling laser industry as much as a belated report from Corning Glass Works in the autumn of 1970 that it had produced a glass fiber pure enough to transport data-laden laser light. That achievement, which we'll study in the next chapter, proved the feasibility of light wave communication and prompted stepped-up research that cleared the way for the use of lasers in medicine and countless other disciplines.

The special qualities of the light beam produced by a laser make it ideally effective on many kinds of human tissue. As an example, the beam of the argon laser is most often used in eye surgery and can be aimed directly through the eyeball without harm. Its energy has no effect until the blue-green beam strikes the red pigment of the retina at the back of the eye, where it is absorbed in an immediate photochemical reaction, causing something of a microscopic explosion in the tissue that rids the patient of his problem swiftly and painlessly.

Other medical procedures require different types of lasers, primarily those using carbon dioxide or synthetic crystals as active materials. The beams of both carbon dioxide and synthetic yttrium aluminum garnet crystal (or YAG) lasers are invisible, and both are generally used for different purposes. Human tissue is about 70 percent water, which easily absorbs the energy of the carbon dioxide laser that has proved to be greatly effective as a cutting instrument, particularly in areas that cannot be reached by a scalpel in a surgeon's hand. The beam of the YAG laser can penetrate human tissue and is excellent for "cooking away" or vaporizing diseased or unwanted tissue. The laser "knife of light" offers other advantages: the risk of infection is reduced because no instrument touches the wound, surrounding tissue is unaffected, and the beam seals off blood vessels to photocoagulate tissue as it cuts, lowering the danger of excess bleeding.

Surgeons use today's fiber-optic endoscopes to pinpoint and eradicate harmful elements with laser beams in nearly every part of the body—including such organs as the lungs, the esophagus, and most of the gastrointestinal tract—without cutting it open. The hope that lasers might be able to abolish cancers has yet to be realized, but doctors have removed some early-stage larynx cancers with brief operating room and office treatments, saving weeks of unpleasant radiation therapy.

The innovative use of lasers in medicine is still in its infancy, despite the fact that more than a million Americans were being treated with lasers by 1988. The bulk of these procedures involve the eyes, but promising new techniques are being discovered regularly in many procedures. The argon laser's propensity for zeroing in on the color red has inspired clinical tests that introduce a red dye, hematoporphyrin derivative, into the bodies of terminal cancer patients. Only the cancerous tissue absorbs the dye, allowing the laser beam to be—theoretically—attracted to the red and in position to destroy the tumor. Testing results will not be known for years, but the potential is there.

Some of the early mundane uses of the laser have since been transformed into lucrative mainstream technologies. Its ability to create holograms has led to the laser's application in a system that ensures the security of credit cards. The technology that produced unpopular videodisc recordings was transferred to the production of compact audio discs, which have been enormously successful. Applications continue to mount as industries explore the laser's special qualities and producers develop lower-priced devices designed for specific tasks. The diversified properties of the laser beam make it a valuable tool for measuring, drilling, cutting, and welding in laboratories, factories, and military and civilian installations of many kinds, even retail stores. The laser made its debut at the local supermarket in the mid-1970s for use in grocery coding and scanning systems and for the marking of expiration dates on various product containers. Its versatility makes it an omnipresent and increasingly valuable product.

In the United States, well over ten thousand people are now working in the laser industry, a $600 million business that generates billions of dollars more in allied fields. The complicated process that produced the laser provided more tribulations and triumphs for two of its key innovators, Gordon Gould and Dr. Ted Maiman. Charles Townes and Arthur Schawlow received most of the credit for breakthroughs leading to the laser and won individual Nobel Prizes for work related to it. Gould spent years and a considerable fortune trying to prove his ownership of laser patents, and Maiman had his own run-in with the Patent Office.

His failure to file a patent before Townes and Schawlow left Gould frustrated until the courts finally granted him the rights to three patents he believed were his: the optically pumped laser in 1977; the process covering drilling, cutting, and welding with lasers in 1979; and the patent for the gas-discharge laser in 1987. All three patents awarded him retroactive rights in addition to the royalties and license fees he was entitled to during the lifetime of the patents. By early 1988, Gould had earned more than $25 million from his patents with much more to come. The long delay actually benefited him financially, despite some $9 million in legal fees, because the patents would have expired if he had obtained them earlier and he would have missed the peak of the laser boom. He sold 80 percent of his expected royalties to two outside investors to finance his legal expenses.

Maiman, convinced his employers "had little idea what to do with the laser," left Hughes in spring 1961 to set up his own laser lab in a smaller company where he hoped he might be better able to profit from his work. He insists the world's first laser cost Hughes a total of $50,000, including—in auto shop jargon—parts, labor, and overhead. Hughes recouped that investment many times over in licensing fees, and, after a first year of uncertainty, other company scientists went on to produce the important Q-switched laser, collaborated in the development of the noble gas ion lasers, and discovered stimulated Raman scattering that led to a versatile technique for optical frequency conversion. Hughes's federal contracts made it a leading producer of laser range finders for the military.

With a major investment from Union Carbide, Maiman reorganized the company he had joined, and incorporated it under the name *KORAD,* to begin the commercial production and marketing of lasers. Several years later, the former Hughes patent attorney informed him that the company had never been able to find the paper he had supposedly signed giving Hughes automatic rights to any inventions he would make during the course of his employment. Maiman believed he had never signed such a paper, meaning Hughes would hold shop rights to the laser but the inventor would own the patent. The Patent Office, in traditional adversarial stance, was now reluctant to award patents to either Hughes or Maiman.

Hughes's long inability to obtain a patent for Maiman's innovation reached a climax in 1967, seven years after he had made the first laser. Maiman's lawyer wrote to the Patent Office asking that he be recognized as co-attorney in the case with the Hughes lawyers—enabling both parties to receive relevant information—and prompted a shocking reply that reeked of bureaucratic incompetency. Only Maiman's attorney received the communication that declared the patent application in final rejection and in danger of being dropped without new supporting evidence. Maiman finally understood the long delay. The Hughes application had not convinced the Patent Office that his laser work had any importance. The bureaucrats described it as *obvious,* citing three key reasons for their thinking: the early Schawlow-Townes paper; Schawlow's rejection of the ruby; and, incredibly, the discredited paper that provided incorrect measurements of ruby fluorescence!

Maiman was understandably furious and wrote a detailed reply that pointed out "that a careful rereading of all the references supplied would prove why a ruby *will not work*—not why it's suitability was *obvious*." The patent attorney believed the letter would win the patent if submitted as an affidavit but cautioned that Maiman now faced an agonizing decision. Acceptance of the affidavit without further action meant Hughes would own the patent and gain all the financial rewards. Maiman could fight on and possibly win the patent—or just let it go forward as it was and settle for nothing more than official recognition as the inventor. Maiman chose the latter path, and Hughes was delighted to submit the convincing affidavit. The patent was finally approved.

Although gaining nothing close to the fortune that Gould is amassing for his contributions, Ted Maiman did quite well in his own business and claims to have no regrets about his decision. He received recognition too late for Nobel Prize consideration but was awarded the two other prestigious international prizes: the Wolfe Prize from Israel in 1984, and the Japan Prize, often called the Nobel of the East, in 1987. Maiman's ultimate recognition as the laser's inventor came with his induction into the Inventors Hall of Fame, a most exclusive collection of esteemed innovators on the level of Edison, Bell, and the Wright brothers.

7

The Fiber-Optics Breakthrough:
The Communications Society

"Popular uses of fiber may break out of something trivial. Fiber services will take off when the average citizen can plug the thing in."

—PETER KEEN, DIRECTOR ICIT

When Robert D. Maurer prepared himself to give a speech before a conference on "Trunk Telecommunications by Guided Waves" in Britain at the end of September 1970, he had a mild case of the jitters that might be experienced by any physicist pulled from the lab to speak before a distinguished body of his peers. He calmed himself with the thought that the audience would agree that he and his colleagues at Corning Glass Works had been engaged in "an interesting bit of work."

At the end of the four-day conference, that work was being hailed as the definitive breakthrough in man's long quest to communicate by using light. What Maurer and associates Donald B. Keck and Peter C. Schultz had done was produce a glass fiber so pure that it had an attenuation of 17 decibels per kilometer at the 632.8-nanometer wavelength of a helium-neon laser. In simpler terms, Maurer and his colleagues had produced a fiber thin as a human hair and transparent enough that if seawater were equally clear, one could float on the surface and minutely examine the ocean floor at its greatest depth. The Corning innovators had surpassed all expectations and created a medium that could transport unprecedented amounts of information on laser beams for commercially viable distances. They had pried open a previously locked door that was sure to lead to a new era in communications and a host of other important disciplines.

Fiber optics is an emerging technology whose potential is far from fully realized, but the *idea* of communicating with light dates back to the signal bonfires of prehistoric tribes. Even Paul Revere's advance warning that "the British are coming" came by way of prearranged lanterns hung in the steeple of Boston's Old North Church, and the French built an "optical telegraph system" two decades later in their own war with the British. That system utilized movable signal arms in a series of high towers spaced at distances that allowed the tower keepers to read the semaphore signals with telescopes and pass along the news.

The nearest ancestor of today's theory of lightwave communication arose in the fertile brain of Alexander Graham Bell. That American communications genius went to his deathbed insisting that something he called the photophone was an even greater invention than his telephone. Bell invented the device immediately after he had won the patent for the telephone and demonstrated it at his Washington laboratory in February 1880. He transmitted the human voice by means of the vibration of a mirror that modulated the intensity of a heightened beam of sunlight and focused it on a selenium rod that was connected in series with a battery at a telephone receiver. Bell demonstrated that sound could be transmitted by sunlight, the brightest light source then available, but atmospheric elements that blocked the sun's rays made transmission through the air impractical. Legend has it that Bell was so excited by his new invention that he nearly named his second daughter Photophone. Eventually forced to abandon the project, he donated his model to the Smithsonian Institution.

Even before Bell was pondering practical methods of sending voice communication by light, an Englishman named John Tyndall was making significant discoveries by exploring the nature of light itself. He succeeded in demonstrating a phenomenon that became known as total internal reflection. Tyndall showed that when light travels through a dense medium such as water and hits the boundary of a less dense medium such as air—at a small angle—it will be totally reflected back into the denser medium. In the case of air and water, the boundary acts exactly as a mirror. Scientists were fascinated by this phenomenon, and Tyndall's discovery kept them exploring the possibilities of light communications for a century. The principles he demonstrated continue to play a role in guiding light through optical fibers today.

Another Englishman of the 1880s, Charles Vernon Boys, also contributed to the early knowledge of fiber-optics technology by creating the first thin glass fibers, although he had no intention of using them to carry light. Hoping to use glass "threads" to replace silk in delicate suspension systems, he devised a method for producing them that was colorfully primitive but effective. He fashioned long strings of glass by attaching molten quartz to an arrow and shooting the arrow from a bow. His method demonstrated that glass could be transformed into fiber that might have a multitude of practical uses.

Despite the stirring of research that created interest in the final decades of the 1800s—including an American patent that proposed to carry light from room to room through pipes with inner reflective surfaces—few sound new theories surfaced until 1934. By then sources of artificial light were strong enough to afford Norman R. French of the American Telephone and Telegraph Corporation (AT&T) a patent for an invention that, on first reading, sounds amazingly close to what eventually was developed. French's patent for an Optical Telephone System called for a network of cables that would carry voice signals on beams of light. The cable might be composed of solid glass or quartz rods, but he saw greater efficiency in a tube with an inner reflecting surface in which the voice signal would be sent through a vacuum rather than air. Two RCA scientists obtained a patent for a similar optical system in the 1940s, but their invention was designed to carry video rather than voice transmissions.

Scientists at Bell Laboratories, following in the founder's footsteps, continued to work on the development of a "light pipe" containing a series of gas lenses that would refract the light and carry it onward. Others in the 1950s, Brian O'Brien of the American Optical Company and a British research group, were demonstrating that practical glass fibers might be more productive. They succeeded in transmitting weak images over bundles of glass fibers they had developed. There was no lack of conviction in the United States or abroad that light impulses sent through some type of transparent medium would eventually carry the human voice and other signals more effectively than the systems then in use. However, the light wave communication theory would remain only that until a vital marriage could be arranged: one between a sound-transparent medium and a powerful, dependable light source.

The second of these problems seemed to be on the way to a solution on June 16, 1960, when Dr. Theodore H. Maiman successfully demonstrated light amplification by the stimulated emission of radiation. As we have seen, Maiman's first laser inspired the subsequent development of smaller and more powerful lasers that were destined to become ideal for light wave communications. Still remaining was the problem of perfecting a medium through which the light could be transmitted with maximum efficacy.

British Post Office researchers had been trying to solve that problem for years; they decided in the mid-1960s to concentrate on sending light through solid glass fibers rather than hollow tubes. Unfortunately, their best optical fiber lost 1,000 decibels of its light power per kilometer, making it virtually useless for communications. New studies by engineers Charles K. Kao and George A. Hockham of Standard Telecommunications Laboratories revealed new information about the possibilities and limitations of optic fiber for communications. They found that impurities in the glass rather than the material itself caused the high losses and concluded that glass fibers would serve the desired purpose only if the impurities could be removed to reduce the attenuation to about 20 decibels per kilometer—a level at which 1 percent of the light would emerge after traveling a full kilometer.

Corning Glass Works, a company most Americans know for its consumer products, entered the glass fiber field in something of a roundabout manner. A Corning scientist, on a visit to England, was told by the British postal people of their unsuccessful attempts to produce optical fibers for communications. He passed the information along on his return, and a new Corning group was formed to explore what America's most established glass works could create to meet the new demand for what was, after all, its corporate "middle name."

The new group was headed by Robert D. Maurer, a physicist, who had no previous experience in working with glass as a communications medium and approached the research with no preconceived ideas. Once briefed on the basic goals of their assignment, Maurer and his associates gathered the smattering of facts available to them and launched the program. Choosing the glass to be used in the hoped-for fiber carried them immediately into unexplored territory. They were aware that scientists in nearly every other laboratory in the world had selected compound glasses as the best for making optical fibers. Compound glasses are composed of a wide variety of glasses that provide many options for achieving the necessary refractive index values and are easier to melt, simplifying the process of transforming them into fibers. The Corning scientists, however, opted for a course that paid greater respect to the basic problem: purity. They selected fused silica, which had the advantage of being highly pure from the start.

It took very little time for the group to discover that their scientific "shortcut" was a possible mistake. Fused silica's high melting point makes it extremely difficult to manipulate into fiber, and it also has the lowest refractive index of almost any of the glasses they could have selected. They needed a glass for the inner core with sufficient refractive index to carry the light and believed the fused silica's purity made it suitable for that purpose. But this meant finding a glass with an even lower refractive index to serve as the outer cladding that could confine the data-carrying light inside the core.

Faced with a problem that would inspire many to

execute an abrupt about-face, the Corning scientists used creative thinking to make the best of what appeared to be a rather unfortunate beginning. Instead of scrapping the idea of a core of fused silica and selecting any number of glasses with higher refractive indices, Maurer and his colleagues decided to use fused silica as the basic material for the entire fiber. The refractive index of their fiber's core would be increased by adding a dopant, leaving the outer part, or cladding, in its normal state.

This approach, seemingly overcomplicated and impractical, would prove to be the most innovative in science's long search for a viable light-guiding glass fiber. Maurer and his associates, Drs. Donald Keck and Peter Schultz, set about to devise their own process for making an optical fiber, beginning with a tube of fused silica that would serve as the outer cladding. Other fused silica—with a variety of experimental dopants added to improve its refractive ability— was deposited inside the tube as "soot" by chemical deposition. The entire glass configuration was then heated, collapsed, and drawn out into a fiber. The process worked, with the doped inner silica having a higher refractive index than that of its outer protective covering. This patented Corning process, called *inside vapor deposition*, later became a standard method for others who produce commercial optical fibers.

The test results for Maurer's earliest fibers were poor, and Corning's commitment to the research was not so high at that time that the group was encouraged to devote full time to it at the expense of other commercially proven projects. The scientists made their best progress during the summer months, when they could afford to take time from their other work. The first fibers to emerge from this innovative method had losses in the range of 1,000 decibels per kilometer. Maurer and associates experimented with other dopants to raise the core's refractive index more effectively and to lower the level of impurities that had caused the losses in their earlier samples. Consistent with Kao and Hockham's pronouncement that an attenuation of 20 decibels per kilometer would permit the use of optical fibers for communication, the Corning team made that level its minimum goal. At long last, using titanium dioxide to raise the core's refractive index, they bettered the minimum mark in 1970. The fiber they created lost only 17 decibels per kilometer at the 632.8-nanometer wavelength of the helium-neon laser used in the testing.

Maurer went to the British Telecommunications Conference unsure of the reception his group's work would receive. To his surprise, most of the papers presented dealt with millimeter waveguides. Corning had made a giant leap forward, and the reaction to Maurer's presentation was vigorous new activity at Corning and around the world. It was clear that Corning would have to improve both product and process to create light-carrying fiber that would be commercially marketable. The 17-decibel loss was a promising starting point, but all efforts would now be directed toward creating a better fiber that could be mass-produced in a factory.

Their review made it clear that refinements were needed in two major areas—a more manageable dopant should be found, and a more compatible core size would simplify the matching of wavelength and light source. Titanium dioxide, the original doping agent, was easy to come by, but the fibers required heat treatment after drawing. The core of the original fiber was only a few micrometers in diameter, making it difficult to insert the light. Lengthy experimentation proved that germanium was a better dopant than titanium dioxide, but the core-size problem remained.

In making the original low-loss fiber, Corning had

invented the inside vapor deposition process in which the core material was deposited inside a start tube that would form the cladding. This time, Corning created a new method called *outside vapor deposition*, or OVD, in which the glass soot is deposited around a mandrel. The mandrel is then removed, and the remaining composite glass preform is heated again and drawn out into a fiber.

The problem of the most utilitarian core size required much contemplation. The original low-loss fiber was in a "single-mode" design, one that could carry only one wave-guide mode of light. Maurer and his group recognized that such a simple design could lead to many problems in practical communications systems.

It would have been relatively simple to make a fiber with a larger core, the multimode step-index fiber that was standard in early attempts to make glass fiber elsewhere. But, although the large core collects light more effectively and allows looser tolerances than the small core, the light pathways have a tendency to scatter, reducing the bandwidth and its capacity for handling large amounts of information. The ability to transmit huge loads of information simultaneously was the value of fiber optics, and Corning dismissed the large-core idea as counterproductive.

A third possibility—graded-index fibers—had been attempted by researchers at the Nippon Sheet Glass Company in 1969. The Japanese used an ion-exchange process to develop a refractive index gradient, but they were not able to reduce attenuation to acceptable levels. The basic theory of graded-index fibers was promising and the Corning group set out to perfect it. Graded-index fibers feature a gradual increase in refractive index from the boundary of the core and claddings to the center of the core. The refraction of the light itself keeps it confined to the core, and, if great care is taken in the graduation of the refractive index, a number of light waves can be sent at similar rates through the fiber. The pulse spreading is thereby reduced and the bandwidth increased.

Corning's Larry L. Carpenter launched experiments that concentrated on depositing a series of glass layers to form the core. Each layer deposited had a progressively lower refractive index, and, in the end, the preform was heated and drawn out into glass fiber. Carpenter was able to produce the first low-loss graded-index fibers in 1972. The introduction of this patented Corning process kept the company ahead in the fiber-optics race despite successes elsewhere. Two Australians introduced a fiber with a liquid core that produced attenuation of only 8 decibels per kilometer, well short of Corning's graded-index fiber with 4-decibel losses.

Within a five-year period, Corning's innovators had produced a fiber that was suitable for telecommunications use. Other firms were also developing optical fibers by the mid-1970s, and other laboratories were working to perfect new lasers capable of sending coherent light through anticipated new communications lines. Lab work would continue in all areas, but the time was approaching when the research and development had to start paying a dividend.

BRINGING THE PRODUCT TO MARKET

Although Corning Glass Works had created the breakthrough that pointed the way to revolutionary changes in communications, it was not equipped to assume leadership in every area of the revolution immediately. The upstate New York firm had joined the search for a suitable optical fiber because *glass* was its business. Other aspects of the emerging technology were not in its range. Decisions had to be made about how the company might best profit from its contributions to light wave communications.

Management knew that, to that point, the communications industry had had little to do with its own glass-related market. Customers and suppliers in communications usually had long-established relationships, and overseas markets were largely controlled by governmental agencies obliged to deal with domestic concerns. In an effort to stimulate interest in fiber optics, Corning briefly manufactured a line of fiber-optic cables in the mid-1970s, but the corporate wisdom was to confine itself to the technology it knew best: glass manufacture. Other companies could take over manufacture of cable and other light wave components.

In its strategy of bringing its product to the overseas market, the company licensed Corning technology to Japanese and Italian firms; formed joint ventures with local companies in France and Britain; and formed a partnership with Siemens of West Germany that became Siecor GmbH. Siecor Corporation was formed with Siemens in the United States to manufacture optical cable.

The laser was the ideal light source for fiber-optic communication, but modulating the good-quality beams provided by gas and crystalline lasers generally required external light modulators that were expensive and prone to operational problems. The semiconductor diode laser was easier to modulate and smaller, making it more suitable for the 50-micrometer diameter of the core of graded-index fibers. The best of them also emitted wavelengths between 800 and 900 nanometers, ideal for transmission through the best fibers available.

But the semiconductor lasers of the early 1970s had flaws that sent scientists to their workbenches. High drive currents necessary for producing laser emission caused overheating and rapid deterioration, and cooling to liquid nitrogen temperatures provided only a temporary solution. Solving this problem became the immediate goal of several research organizations. Laser Diode Laboratories Inc., which later became M/A Com Laser Diode Labs, commercially introduced semiconductor lasers in 1975 that were capable of operating continuously at room temperatures for one thousand hours, and two years later Bell Labs demonstrated semiconductor lasers that were capable of continuous room-temperature operation for more than 1 million hours—or a hundred years.

Research was progressing on the incoherent light-emitting diode (LED), which would be especially useful in short-distance or limited bandwidth transmission. LEDs, structurally similar to semiconductor lasers, produce lower optical power and broader beams than lasers, but they are long-lived and easier and less costly to manufacture. Silicon detectors that are sensitive at 800 to 900 nanometers and capable of transforming optical signals back into electrical form were already available.

Initial worries about the ability to connect glass fibers produced two separate new systems that proved workable. Fibers used in telephone links and other permanent installations could be joined by melting the ends together and forming a weld when they cooled. The Department of Defense took the lead in developing fiber-optic analogues of electrical connectors that could be used for changeable connections in military equipment, and a number of companies began producing connectors that met the demand for use in computers and other instruments.

Two communications giants, AT&T and General Telephone (GTE), vied to become the first to test fiber-optic telecommunications for commercial use. That race ended in a near dead heat in spring 1977 with GTE's prototype going into operation a bit sooner, and AT&T's field test providing more ambitious results. AT&T's system demonstrated its ability to carry the customer's voice, video, and data communications traffic over two lengths of about a kilometer each in Chicago at a rate of 45 megabits per second—corresponding to 672 voice channels. GTE used two repeaters (amplifiers that strengthen weak signals) to cover 9 kilometers in Long Beach, California, at the less demanding rate of 1.5 megabits per second. Corning supplied fiber for both operations.

The tests were closely monitored for the better part of two years by everyone interested in the future of light wave communication, and the reports were uniformly positive. Both systems were performing up to the expected standards, and the trade press began to rail against the telephone industry's slowness in taking full advantage of the exciting new technology. However, behind the scenes, positive results from Chicago and Long Beach inspired a great deal of quiet action. Postal and telephone authorities in other parts of the world and other independent U.S. telephone companies began to set up tests of their own fiber-optic systems.

From Corning's viewpoint, the successful tests indicated that large quantities of optical fiber would be needed soon. The company quickly got a pilot plant operating in Corning, New York, and also began work on a full-scale fiber manufacturing facility in Wilmington, North Carolina. AT&T's manufacturing arm, Western Electric, had negotiated a patent license for the Corning fiber process in 1970 and 1975, and its Chicago success spurred plans for developing plants to make its own fiber and communications systems.

Farsighted governmental agencies in Japan and Canada were the first to explore the more remarkable properties of light wave communications. Japan's Ministry for International Trade and Industry announced plans to build a two-way system called Hi-OVIS (highly interactive optical visual information system) a year before the limited Bell and GTE tests began. The innovative system began serving 150 homes in a Japanese town in July 1978, providing subscribers with access to a wide range of information as well as the ability to originate programs from their own homes for transmission over an interactive video network. Japan saw the adventurous undertaking as a satisfactory demonstration of the future validity of the new technology, despite its high cost and the rapid obsolescence of its expensive equipment.

Canada began planning a similar but less ambitious system at about the time Hi-OVIS started operating in Japan. The Canadian program—under the joint sponsorship of the Department of Communications, the Canadian Telecommunications Carrier Association, the Manitoba Telephone System, and Northern Telecom—was designed to test the potential of fiber optics for telecommunications in the small, isolated towns of Elie and Ste. Eustache in rural Manitoba. The existing telephone systems were primitive, with four or more families sharing party lines, and the Manitoba demonstration would also determine whether glass cables could withstand winter temperatures of 40 degrees below zero. The program was less expensive than its Japanese forerunner, in part because it was not interactive, but the finished system gave villagers single-party phone lines, cable television, stereophonic FM radio, and the Telidon videotex service, which provides a wide range of information.

Even while these early commercial applications of fiber-optic systems were being planned and implemented during the closing years of the 1970s researchers in laboratories around the world were continuing to look ahead, envisioning improved optical fibers and the new light sources that would complement them. The 1970s had been a decade of breakthrough and development. Many believed the 1980s would be a decade of rapid deployment and financial boom.

INNOVATION AND EVOLUTION IN A NEW TECHNOLOGY

The rush to expand the technology that Corning had pioneered allowed its innovators no time to rest on their laurels. Corning's roots in the glass industry go back to 1851, and the company gave early recognition to the value of innovation by establishing its own research laboratory in 1908. Over the past hundred years Corning has produced three momentous innovations that have revolutionized the lives of millions: the production of the first incandescent lamp bulb enclosures for Thomas A. Edison, the development of the all-glass television bulb, and the invention of commercially feasible optical fibers for telecommunications. But its innovative optical fiber research and development program faced strong competition in the late 1970s from scientists elsewhere—particularly Japan.

The first generation of fiber-optic systems all operated at a wavelength range of 800 to 900 nanometers. Scientists soon theorized that scattering losses inherent in silica fibers would diminish as the wavelength grew, an important consideration in planning for long-haul telecommunications because lower losses mean signals could travel greater distances between repeaters. The variation of the refractive index of glass with wavelength causes a phenomenon known as *wavelength dispersion*, in which light of some wavelengths travels faster than that of others, causing a pulse to spread out. Wavelength dispersion can limit bandwidth or the amount of information that can be transmitted by a signal.

It was believed that nature provided the least wavelength dispersion—and an exceptionally large bandwidth—at 1,300 nanometers.

Longer-wavelength research progressed rapidly with a key breakthrough in 1976 coming from Japan, where fibers that registered losses of only 0.5 decibel per kilometer at 1,300 nanometers were made. Scientists in both the United States and Britain measured zero material dispersion in 1,300-nanometer fibers the following year. Semiconductor lasers of indium gallium arsenide phosphide with lifetimes of 1,500 hours at the 1,300-nanometer wavelength were successfully demonstrated in 1977 by J. J. Hsieh and his associates at MIT's Lincoln Laboratory. The longer wavelength also required new detectors to replace those of silicon that were insensitive at 1,300 nanometers, but the development of germanium detectors and compound semiconductor devices made of indium gallium arsenide families quickly solved the problem.

Scientists at Corning and elsewhere were aware that wavelengths of 1,550 nanometers, or 1.55 micrometers, also held great promise and began programs to perfect those fibers. Silica fibers would transmit ever-diminishing light beyond 1.55 micrometers but would have their absolute lowest loss at that precise wavelength. Researchers from the Nippon Telegraph and Telephone Corporation recorded the first success at the new wavelength with single-mode fibers that lost only 0.2 decibel per kilometer at 1.55 micrometers. They also predicted, in a paper presented in 1979, that losses of 0.18 decibel per kilometer would be the absolute limit in that wavelength.

Early in 1982 Corning made a single-mode fiber with an attenuation rate of only 0.16 decibel per kilometer at 1.55 micrometers, meaning 1 percent of the light put into the fiber would remain after traveling a distance of 77.5

miles. The major problem with that wavelength was the reappearance of dispersion in standard fibers that didn't exist at 1,300 nanometers, but later research demonstrated that it was possible to make fibers with low dispersion throughout the range between 1.3 and 1.55 micrometers by carefully controlling the fiber composition.

EXISTING APPLICATION VERSUS LONG-TERM PROMISE

Two decades have passed since Robert Maurer and his Corning colleagues produced the first optical fiber that could effectively carry light. Developments in innovative laboratories around the world have been rapid and startling. But the overriding question about this mind-boggling new technology remains: Has the practical application kept pace with the long-term promise?

In Robert Maurer's opinion it hasn't. He feels fiber optics is destined to "take over almost all of communications. Maurer, recipient of the 1987 John Tyndall Award for his contribution to fiber optics, says, "Optical fiber will improve our lives as much as any existing technology, once these and other problems are solved."

Fiber optics has been used best by the telecommunications companies, where the advantages are readily apparent. Light guide cables are considerably thinner and lighter than their conventional copper counterparts, enabling companies to avoid construction costs by installing them inside existing ducts carrying standard cables. Glass fiber's inability to conduct electricity provides interference-free reception and, although the digital pulses of copper transmissions must be regenerated after traveling about a mile, light waves have been sent 100 miles or farther before they needed restrengthening. Today's most advanced light wave systems can relay data at 1.7 gigabits per second—fast enough to transmit the entire Encyclopaedia Britannica in just *two seconds*.

With the court-ordered breakup of AT&T, several telephone companies began offering light wave long-distance service, and more than a dozen have since undertaken the construction of an additional total of 7 billion circuit miles of fiber-optic lines. AT&T's 1,500-kilometer system between Boston and Richmond, Virginia, was put into service in 1984, and the new technology received more public recognition that same year, when AT&T's FT3C network in Los Angeles was used to carry live coverage from the various sites of the Olympic Games to the television broadcast center. Since then, AT&T, MCI Telecommunications, and other companies have installed systems longer than the milestone Northeast Corridor.

Optical fiber received a big boost as the new communications cable of choice in 1984 when AT&T and twenty-eight other telecommunications administrations from North America and Europe selected it instead of coaxial cable for use in the eighth transatlantic undersea cable system that began operation in 1988. TAT-8 can transmit the equivalent of 37,800 simultaneous telephone conversations as well as the equivalent combination of voice, data, and video signals across 3,607 nautical miles of the ocean floor. The system utilizes two working pairs of single-mode 1,300-nanometer fibers, each driven by 1.3-micron lasers operating at 296 megabits per second. Through light guide technology, branching of undersea cable, allowing landing points to be used in Britain and France, is possible for the first time, and only 125 repeaters are needed to span the Atlantic—far fewer than were necessary in earlier cables. A similar trans-pacific system, using 250 repeaters, is also planned.

In 1986, two years after announcing the undersea light wave systems for the Atlantic and Pacific, AT&T, Teleglobe Canada, British Telecommunications International, the French Secretariat of State for Post and Telecommunications, and Spain's Telefónica began planning the second undersea light wave link between Europe and America. TAT-9, expected to be ready in 1991, will benefit from new advances and operate at 565 megabits per second, nearly double the capacity of TAT-8. The 1991 undersea system will employ a unique submerged multiplex branching repeater, permitting it to connect the United States and Canada with separate points in Britain, France, and Spain.

Early 1988 saw enough optical fiber already installed in North America to bring nearly unlimited information capacity to within a very few miles of 80 percent of the total population, but the inability to transport that cable on into the homes of individual subscribers has created boundless frustration for the technology's earliest supporters. They envisioned "fibered" homes, apartment buildings, and *cities*, knowing that the ultimate success of any revolutionary scientific advance—the telephone, household electrification, radio, television—depends on its appeal to the general public. The United States with its affluent, educated, information-mad millions was expected to lead the way. Curiously, this has not happened in the nation where Maurer and his colleagues gave birth to fiber-optic communication.

Recent years have brought encouraging developments in the United States on an institutional level. More and more office buildings, college campuses, hospitals, and the like, have installed fiber-optic systems to upgrade their overall communications capacities. And there has been a definite trend toward installing broadband capacity in *new* real estate developments: commercial buildings, office parks, and a few recently constructed housing subdivisions. But Japan and France are ahead in bringing the marvels of light wave communication capabilities to the average consumer.

Japan, which pointed the way with its Hi-OVIS system in a single village in 1978, has continued to expand and build on its services. Some Japanese telephone subscribers can now buy a system called Teleface—consisting of a highly sensitive television camera and a black-and-white monitor—that can be plugged into the telephone to provide face-to-face conversations with those on the other end of the fiber-optic line. The state-owned French telephone company connected 1,500 homes in the resort city of Biarritz to a fiber-optic network that provided consumers with videophone service, videotext, and cable television in 1983. The service became an instant success as the world's first mass consumer videotext system with more than 2.5 million French households equipped with terminals by 1986. The Minitel system continues to expand and flourish with governmental financing.

To this point, there have been no equivalent successes in bringing fiber-optic advances into the homes of American customers. Most analysts say that economic factors, especially the high costs of the electronic end equipment, are primarily at fault. When new expertise enables prices to be lowered and the public is made fully aware of the myriad services these systems can provide, there is every reason to expect that fiber optics will engender monumental changes in the way Americans conduct their personal and business lives. They will have the freedom to do everything from shopping to banking to research to teleconferencing in their own homes. Many think that new options for home entertainment such as digital television sets and the much-wider choice of channels through fiber cables—plus interactive fun and games as yet unimagined—will eventually spur demand.

Even more exciting to the scientists working to perfect light wave technology is the promise it holds for improving the performances of a wide range of relatively new technology. These innovators see fiber optics transporting us from the Electronics Age into the Age of Photonics. They say that photonics is functionally about where electronics was several decades ago, but is poised for a great leap forward when electronic devices are no longer needed to control light wave signals. Scientists in a number of laboratories are attempting to design optical equivalents of transistors—devices to control light by light. Once these photonic controls are perfected, the path will be cleared for such fiber-optic innovations as the much-anticipated photonic computer, with parallelism and speed at least one thousand times greater than today's best.

Avoiding the additional expense of converting light pulses into electronic form and back again will permit cost reductions that will encourage expansion of the marketplace. Corning's great work in producing glass fibers through which light signals can be successfully transmitted has inspired major complementary innovations that have enabled transmission capacity to double annually. Impressive progress, no doubt, but with fiber optics, the very best is yet to come.

8

Mevacor:
Developing the Weapon
Against Cholesterol

*"We try never to forget that medicine is for the people. It is
not for the profit. The profits follow, and if we have remembered that, they have never failed to appear."*
                                                    —GEORGE W. MERCK

As outlined in our study of cyclosporine in chapter 3, the
biomedical sciences have had to struggle to keep pace with
recent innovations in other areas that are allowing men to
do many things with a speed and ease undreamed of in the
past. Discovering agents that might protect or prolong
human life has always been an agonizingly slow process.
Merck & Company, the Rahway, New Jersey, pharmaceutical giant, was one of many organizations that spent
decades searching for a drug that would control cholesterol.
Ultimately, after investing much time and effort in devising

innovative methods that might solve the mystery, Merck's classic breakthrough came with astonishing quickness.

Scientists have believed for nearly a century that the mysterious substance called cholesterol was a major contributor to atherosclerosis, a disease that clogs the circulatory system with dangerous fatty deposits. Close to 5.5 million Americans have been diagnosed as victims of symptomatic coronary heart disease and more than half a million people die from it in the United States each year. The number-one cause of death, it takes more lives annually than all forms of cancer combined. Economic cost in the United States alone is estimated at $60 billion annually.

The unraveling of the mystery of how cholesterol is formed and how it might be inhibited was a step-by-step process with contributions from researchers in every corner of the globe. Merck scientists found a number of the missing pieces of the medical puzzle and delivered the key breakthrough in November 1978 with the discovery of an inhibitor called lovastatin. Almost nine years later, in September 1987, Merck was finally authorized to bring lovastatin to the market under the brand name Mevacor.

A brief rundown of the known history of cholesterol reveals the monumental task science faced in attempting to control it. It began in 1789 when Parisian chemist A. F. Fourcroy first isolated a white fatty substance from gallstones, but it did not even have a name until another Frenchman, M. Berthelot, identified the compound as an alcohol in 1853 and called it cholesterol. Its destructive nature was first recognized in 1904 by a Leipzig pathologist who identified the disease named atherosclerosis, and the first major research breakthrough came in czarist Russia from 1908 until 1913. The Russians demonstrated that a diet of milk and egg yolks induces atherosclerosis in rabbits and that adding pure cholesterol to the diet also induces the disease.

German scientist H. O. Wieland won the Nobel Prize for Chemistry in 1927 for his research on the structure of cholesterol, but little new progress was made until the remarkable development of radioactive isotopic labeling shortly before America's entry into World War II. Scientists were finally able to begin charting the complex manner in which the body produces lipids. Dr. Konrad Bloch made the first major contribution early in the 1940s with his discovery that the presence of acetate always precedes the production of cholesterol. And, in 1951, Dr. Feodor Lynen managed to isolate active acetate, demonstrating that it was an ester of coenzyme A. In recognition of their efforts, Bloch and Lynen won the Nobel Prize for Medicine in 1964.

In 1956, Dr. Karl Folkers, Dr. Carl Hoffman, and others at Merck Sharp & Dohme Research Laboratories identified a substance called *mevalonic acid*, and Dr. Jesse Huff quickly proved that it was another element in the production of cholesterol. Later research elsewhere defined the substance better, and Merck's discovery of mevalonic acid would come to be recognized as the missing piece of the scientific puzzle of cholesterol synthesis. Merck, still unaware of the full significance of its discovery, began an innovative line of experimentation in the late 1950s.

Knowing that cholesterol played a vital role in the formation in the liver of bile acids, which were then secreted into the intestine and almost completely reabsorbed, the Merck people set about finding a method to prevent the natural reabsorption process. The theory was that blood cholesterol levels might be significantly reduced because of the liver's need to "raid" the bloodstream to obtain the cholesterol necessary for making bile acids. The key to success was finding a resin that had a propensity for effectively binding itself to the bile acids in the intestine. Once bound, the resin could then carry the acid and harmful cholesterol on out of the body.

More than one hundred different resins were considered before a synthetic called *cholestyramine* was successfully developed. Testing proved that cholestyramine was relatively effective, but Merck never marketed it as a cholesterol-lowering agent. Produced in a powder form that had to be suspended in liquid, it had a texture and taste unpleasant to many patients. Constipation and other intestinal discomforts were also evident in some test patients. Merck later licensed the product to Mead-Johnson, which put the substance into a more palatable form, changed the name to Questran, and made it a success.

While Merck's scientists were focusing their attention on cholestyramine, others were making important new findings about the role of the enzyme HMG-CoA reductase in the biosynthesis of cholesterol. Dr. Feodor Lynen in Germany demonstrated that HMG-CoA reductase acted as a catalyst that transformed HMG-CoA into the mevalonic acid discovered earlier at Merck. HMG-CoA reductase appeared to have the ability to be a major rate-limiting step in the body's production of cholesterol.

With modern technology making available more sophisticated equipment in the 1970s, scientists in both industrial and academic laboratories initiated fundamental changes in their quest for knowledge about cholesterol synthesis. Biochemistry and use of cell-structure screening, enzymatic pathways, and other such techniques began to receive increased attention. Among the standout academic scientists in the field were Dr. P. Roy Vagelos and Alfred Alberts of Washington University School of Medicine in St. Louis. Vagelos had been a young resident physician before joining Dr. Earl Stadtman, one of the world's leading enzymologists, in a study of the body's methods of producing fatty acids at the National Heart Institute (NHI) in Bethesda, Maryland, in 1959. Vagelos hired Alberts as his assistant and headed the NHI program before both moved on to St. Louis in 1966.

Vagelos and Alberts shed important new light on the mechanics of lipid production and the manner in which biological systems regulate the rate of lipid synthesis. They discovered that animal cells actually needed cholesterol to maintain their normal membrane structure, and a cell with a single enzyme defect lost its viability along with its ability to manufacture cholesterol. All cells of the body required some cholesterol in their membranes, but, once resupplied with cholesterol, cells returned to normality. In theory, then, the synthesis of the lipid could be turned off by regulating the normal function of any one of the more than twenty-five enzymes involved in the overall process. Others later reported similar findings. HMG-CoA reductase was clearly becoming the most likely target in the enzymatic chain.

The scientific community took special note in 1973 of new research being conducted by Drs. Michael S. Brown and Joseph L. Goldstein at the University of Texas Health Science Center in Dallas. Their new study centered on an inherited condition known as familial hypercholesterolemia. Levels of low-density lipoproteins (LDLs), particles that circulate in the blood and carry most of the body's cholesterol, are high enough in these victims to induce heart disease in childhood or the very early adult years. These patients, Brown and Goldstein learned, had unusually high levels of HMG-CoA reductase and cells lacking specific receptors that attract low-density lipoproteins, enabling the cell to absorb it. Unable to recover LDL from the circulation, the cells of the victims respond to that deficiency by accelerating the biosynthetic process to make more, leaving ultrahigh levels in the bloodstream. Their discovery of LDL receptors brought Brown and Goldstein the 1985 Nobel Prize for Medicine.

The rush of new knowledge would enable someone to make the final breakthrough in the search for an effective weapon against cholesterol, and the people at Merck were determined that their company was not going to finish second in the race.

THE PATHWAY TO MEVACOR

A twelve-year heart study in Finland showed that a diet of reduced saturated fat and cholesterol along with a heavier reliance on polyunsaturated fat resulted in a 53 percent decrease in death from heart disease in men and a 34 percent reduction in women. It was a strong boost for the lipid hypothesis, but too few patients were studied to make the results definitive. The best of diets could not always do the job alone, because dietary intake contributes only about a third of the body's cholesterol. The rest is produced by the body in a multistep process.

The need to come up with innovative methods of tracking down an HMG-CoA reductase inhibitor prompted Merck's management to consider changes in the company's research structure. Convinced that the discovery of a potent anticholesterol agent was imminent, Merck sought advice from Dr. Roy Vagelos of Washington University. Vagelos agreed to act as a consultant while the company set up its own culture assay in 1974, and his work impressed Merck to the point that he was invited to become senior vice president in charge of research the following year. Vagelos accepted and brought longtime associate Al Alberts along to head the effort to find the elusive cholesterol fighter.

157

Vegalos and Alberts respected the traditional system of chemical synthesis but saw even more potential in the alternative approach—the screening of microorganisms that could lead them directly to a natural inhibitor of HMG-CoA reductase. That same approach would allow researchers to narrow the areas of concentration and speed the process of other drug development. Alberts spent his first three years at Merck setting up the laboratory, establishing new approaches, and generally creating conditions that he and his mentor hoped might lead to new drug discoveries.

While Alberts and his Merck colleagues were engaged in this seemingly unproductive travail, Japan's Sankyo, Inc., pharmaceutical company made the first significant breakthrough in 1976 with the isolation of a compound called *compactin*. Using techniques similar to those being set up at Merck, a researcher named Akira Endo spent many months looking for an inhibitor in natural microorganisms, such as the one that led to the discovery of penicillin years earlier. Endo and associates tested well over six thousand compounds until a soil mold called *Penicillium citrinum* demonstrated that it could inhibit HMG-CoA reductase and function in vivo to lower levels of cholesterol in the blood. Sankyo patented its discovery in 1974, and articles describing it began to appear in 1976, providing a good lead for researchers everywhere.

At about the same time, Beecham Laboratories in Britain made an almost identical discovery, not even attempting to find an HMG-CoA reductase inhibitor. Beecham published additional details of its research with microorganisms, and the helpful information from both quarters inspired the Merck team to redouble its efforts. Demonstrations that viable inhibitors were present in soil molds bolstered Al Alberts's faith in the correctness of his own approach. The development of antibiotics had made it clear that more than one effective compound could be found in nature.

It was not until late September 1978 that Alberts and his associates felt ready for a final testing phase that would give them the confidence to move into the actual realm of the unknown—the testing of exotic microbial broths that might yield an HMG-CoA reductase inhibitor—without destroying the new systems. Alberts went to Merck's New Lead Discovery Department, under the direction of Dr. Arthur Patchett, and asked for a supply of one hundred microbial broths—old ones that had already been tested by other groups for other purposes. Alberts could validate the new system by *not* finding a variety of compounds that seemed to inhibit HMG-CoA reductase.

It took him and associate Julie Chen two or three days to test the first one hundred broths, and they were delighted to find that nothing unexpected happened to any of the extracts. They were not destroying the enzyme, and the system was working in the face of the unknown cell "garbage." Alberts then asked to be put on Patchett's list of those who would routinely receive supplies of new microbial broths. The department was supplying fifteen to twenty of the new extracts per week for a variety of research groups in the laboratory.

Alberts and Chen received a total of seventeen of the mysterious broths during their first week of serious testing in November. Thanks to their preparation, actually identifying an effective inhibitor should be relatively simple. Chen would prepare body-temperature mixtures of *radioactive* HMG-CoA and HMG-CoA reductase in thin test tubes. Then, using special devices developed for the job, she would put into each tube a different microbial extract. If the tested

159

extract was an effective inhibitor, the enzyme would not be transformed into mevalonic acid and nearly all traces of radioactivity would disappear. Not surprisingly, none of the first week's broths prevented formation of mevalonic acid.

Julie Chen usually arrived at the lab later than Alberts after getting her two children safely off to school and stayed later in the evening to make up for lost time. It took her more than an hour to prepare a series of tubes containing a variety of the new extracts that Patchett's department had sent them. Sometime after midmorning on the first day of the second week of testing, Alberts was a bit surprised to see Chen standing in the doorway of his office across the hallway from the lab. His orders were that he must be informed immediately of any unusual results. Chen tried to suppress her excitement as she told her boss that the very first tube she tested, Two-1 (second week, first sample), showed *no activity*. No radioactivity was present, meaning that no mevalonic acid had been formed!

When interviewed, Al Alberts still demonstrated the wonder and excitement he felt on that special November morning.

There were several explanations of why this had happened. One of them, of course—and in my mind the least likely at the time—was that there was an inhibitor there. The most likely explanation was that, by mistake, the enzyme had been left out of the tube. Without the enzyme, you wouldn't get a reaction. So that's what went through my mind. I probably said something like "That's very nice but—" I mean, I couldn't believe it at all. I couldn't *believe* it—but, boy, I was sure hoping!

You know, when you work in the lab—

working in the unknown—the one thing you have to have, all the time, is hope. Your success rate, by definition, is going to be very, very low. So when you do see something, even when you're almost positive that this is some kind of mistake, there's still the hope. So maybe you go home at night and find yourself saying, "My God, maybe this is right!" The first thing we did, of course, was retest. So, the next day, Julie set it up again and we went through the whole process once more. And, lo and behold, there it was. There was our—*Eureka!*

Alberts is the first to admit that he was dumbfounded by their ability to find the HMG-CoA reductase inhibitor so quickly—in the first test tube of microbial broth in the second week of testing—actually the *eighteenth* sample tested. He is even quicker to point out that what really made the discovery seem so fast and simple was the years spent setting up the system and understanding what they were working with, making the system supersensitive to detect and help them find things more easily, if they were there.

"I hate to say this now," he chuckles, "but after making the system so sensitive so we could find very, very small amounts, what we found was a large amount right away. We actually could have found it with a much less sensitive assay. But I certainly don't regret doing it."

The first order of business was to determine whether Merck's scientists had only "rediscovered the wheel," as Alberts puts it. Had they merely come across the same compound produced earlier by Sankyo and Beecham? Finding the answer to that and other vital questions took four full months of work by scientists in a number of departments. It was first necessary to grow the culture again, in a variety of conditions, to determine whether it would consistently produce the same results. Buoyed by the potential of the new discovery, then called *mevinolin*, Merck laboratories bustled with activity in the weeks that followed.

Dr. Carl Hoffman, who had been instrumental in the earlier discovery of mevalonic acid, managed to isolate the pure inhibitor in a few weeks. A team of spectroscopists headed by Dr. Georg Albers-Schonberg succeeded in determining the inhibitor's precise structure in a matter of days. Alberts, meanwhile, had to prove that his discovery could not only block the enzyme but also block actual cholesterol synthesis. Using methods he had employed at Washington University, Alberts's extensive testing demonstrated that *Aspergillus terreus* limited the production of cholesterol at very low concentrations. In time, it became clear that mevinolin was a *new* and most potent discovery.

Al Alberts remembers with pleasure the memo he sent to Roy Vagelos, then president of the research laboratories, the man with whom he had worked on lipid research for many years. "This memo was really loaded and I think I even wrote on it somewhere, 'Please answer.' Well the memo came back to me about twenty-four hours later with a scribbled note that said, 'Have you purified it yet?' That was a joke, of course—his way of saying go to it! But what he did was get everybody up and running!"

Merck's scientists put in long hours studying mevinolin and the best methods of retrieving it from its microbial broth. Key breakthroughs occurred between late spring and midsummer when Patchett's New Leads Department developed better techniques for isolating the pure inhibitor at about the same time Drs. Richard Monaghan and Edward Stapley were devising improved fermentation methods. Those developments allowed Merck to produce sufficient quantities of mevinolin to begin testing on animals.

Dr. Patchett credits a stroke of good fortune, along with some good work by Dr. Monaghan, as the reason the culture that produced mevinolin was available for Chen and Alberts. It had been grown in Merck's laboratories in Spain in an effort to control protozoa that infect poultry and reduce their growth rate. Patchett had the microbiologist rig up an assay in bacteria, but the culture did not show the same activity in Rahway that it had showed in Spain. Normally, it would have been discarded.

"But Dick Monaghan," Patchett says, "had the good luck or good sense to put it into our FRPS [fermentation products for screening] system. He knew we wanted interesting-looking cultures to go into our FRPS program and, in his judgment, this was an interesting fungus. So he kept it, and we eventually sent samples of it off to a number of people, including Al Alberts."

Mevinolin demonstrated a clear and dramatic effect in reducing plasma cholesterol levels in the animals tested. Confidence soared as preliminary studies indicated that mevinolin was two to three times more potent than compactin. Merck was awarded a patent for mevinolin in 1980, the same year Sankyo of Japan received its patent for Monacolin K, a drug derived from a different organism that had a chemical structure almost identical to that of mevinolin.

New reports in 1980 again pointed out the need for an anticholesterol weapon. One of them was a long-term study involving thousands that had begun in Framingham, Massachusetts, in 1949. That large scientific sampling verified that individuals with very high blood cholesterol levels were far more susceptible to heart attacks than those with levels below 200 miligrams per deciliter (mg/dl). It also confirmed that elevated cholesterol was as great a risk factor for heart disease as smoking and hypertension.

That same year Dr. Ancel Keys of the University of

Minnesota published his noted Seven Countries Study that found wide variances in cholesterol levels in men of different countries and showed that rates of heart disease clearly increased with the levels of cholesterol. As an example, a ten-year study of men from a Finnish village where the mean cholesterol level was 265 mg/dl showed the incidence of fatal heart attacks fourteen times higher than that of men in villages in Japan and Yugoslavia, who had average cholesterol levels of 160 mg/dl.

The need for an effective cholesterol fighter was never more apparent, and the scientists at Merck Sharp & Dohme Research Laboratories were confident they had found it.

THE LONG QUEST FOR APPROVAL

Innovation and inventions in other fields often can be taken to the marketplace as soon as sufficient financing, deployment, and/or production can be arranged. This is definitely not the case with pharmaceuticals, especially in the United States. Because of the possible danger to human health and life, medications of all types—internal and external—must undergo long periods of meticulous testing before the U.S. Food and Drug Administration (FDA) allows them official entry into the commercial marketplace. This same scenario was followed in the case of Merck's mevinolin.

The tests on animals that started in mid-1979 had been enormously successful with almost no untoward side effects noted, and Merck was able to start testing with human volunteers in April 1980. The carefully monitored clinical testing, under the direction of Dr. Jonathan Tobert, proved quickly that mevinolin could produce spectacular reductions in blood cholesterol levels with no significant side effects.

But something happened on the far side of the globe that prompted Merck to halt human testing less than six months after it had begun. Rumors began circulating that Japanese researchers were having problems with compactin, the HMG-CoA reductase inhibitor discovered by Dr. Endo in 1976, and its testing was abruptly halted. The most alarming of the rumors, still neither confirmed nor denied by Sankyo, made the company end its compactin studies because its researchers had discovered intestinal tumors in tested dogs. None of Merck's studies on animals or humans had led to anything of that nature, but the company voluntarily suspended its clinical testing of mevinolin in September 1980.

"Black September of 1980," Tobert says. "I can well remember that occasion. One learns to be philosophical about such things in this business, but still, when you're going full speed like that, it's rather like a runner leading the race and all of a sudden he trips and falls."

For the better part of a year and a half Merck scientists concentrated on long-term laboratory studies and animal testing as a contingency approach to its anticholesterol research until urgent requests from the medical community brought them back into human testing in 1982. Prominent clinicians Dr. Roger Illingworth of Portland, Oregon, and Drs. Scott Grundy and David Bilheimer of Dallas, Texas, associates of Brown and Goldstein, had hypercholesterolemia patients at high risk of dying of heart attacks unless mevinolin could be used to save them. Armed with these humanitarian requests, Merck received FDA consent to make mevinolin available in these cases. Therapy began, and the results were impressive.

Reports on the success with the hypercholesterolemia patients in Oregon and Texas brought new backing that enabled Merck to start planning the resumption of its formal mevinolin testing in 1983. Dr. Daniel Steinberg of the University of California at San Diego and Dr. Jean Wilson of the University of Texas of Dallas, experts on metabolic disorders, strenuously urged a speeding-up of the mevinolin testing program on behalf of the countless people with dangerous levels of blood cholesterol who were doomed to die of heart disease unless those levels could be lowered. And the formal publication of a complex study by the National Heart Lung and Blood Institute also greatly added to the new momentum that would encourage Merck to resume full-scale clinical testing.

That organization's lipid research clinic had conducted a long multicenter, randomized, double-blind placebo-controlled study involving nearly four thousand American men to determine, once and for all, the validity of the lipid hypothesis. Ironically, the tool used to lower cholesterol was cholestyramine, the earlier Merck discovery that had been dropped in favor of the program that eventually produced mevinolin. Released in 1984, the study proved that high cholesterol levels were directly related to coronary artery disease and that the chance of having a heart attack could be reduced by about 2 percentage points for every percentage point by which cholesterol was lowered in the bloodstream. The use of cholestyramine afforded a 19 percent lower rate of heart attack than dietary measures alone. With preliminary evidence that mevinolin was far more effective than cholestyramine, Dr. Tobert launched Merck's full-scale clinical testing of the new agent in May 1984.

Mevinolin clinical testing began producing dramatic results. All of those admitted to the clinical test programs were voluntary patients with life-threatening levels of blood cholesterol that other measures were unable to correct. Dr. Tobert and his staff designed the studies with advice from doctors at the academic clinical centers where the therapy took place. The performance of mevinolin was so astounding that it sometimes led to confusion, particularly for one technician at a Cincinnati medical school.

"The technician who was running the cholesterol analyses," Tobert remembers, "kept repeating them because he couldn't believe what he was seeing. He had seen a patient come in with a base line of about 300 and then come back four weeks later, having been on mevinolin, and instead of 300 it would be 180. He'd never seen anything like this before so he kept repeating the test. He thought he'd somehow messed up the assay—or there must be something wrong with his machine."

Many of the stories from the period took on the aura of medical miracles: An emergency operation to open a blocked carotid artery, the major blood vessel in the neck that feeds the brain, was needed to save the life of a thirty-nine-year-old insurance executive, but mevinolin treatment lowered her blood cholesterol level of 475 by 54 percent; a forty-seven-year-old Eastman Kodak employee endured lifesaving triple bypass surgery before being admitted to the mevinolin program that brought his cholesterol level down to a healthy 175 from a previous count of 300; a thirty-eight-year-old bank manager suffering from hypercholesterolemia had to undergo a twice-monthly procedure called plasmapheresis that removed all the blood from his body, cleansed and reinjected it, as the most effective method of lowering cholesterol and preventing a second, possibly fatal, heart attack. Mevinolin freed him of the plasmapheresis and lowered his cholesterol count to safer levels.

The most heartwarming story to emerge from the clinical studies of mevinolin concerned the plight of a Texas girl who was suffering from severe hypercholesterolemia. It was given prominent space in the pages of *The Journal of the American Medical Association* and in Merck's company magazine. She suffered a near-fatal heart attack at the age of five, and, in her sixth year, doctors performed double bypass surgery and replaced a heart valve in an attempt to keep her alive. But neither the surgery nor a rigid diet reduced the fatty deposits that were increasingly blocking the vital blood supply that carried nourishment to muscle and organ tissues. The child's heart and liver were edging ever closer to complete failure.

In desperation, her physicians opted for heart and liver transplants as their patient's best chance for survival. Donor organs were found, the operations were performed, and for a short time the little girl's condition improved. The new liver was better able to remove harmful deposits from the blood, but, within six months, the cholesterol levels had again risen dangerously high. Having heard of the mevinolin testing, the child's doctors asked for permission to use the new drug as a last-resort measure for their young patient. Permission was granted, and the results were almost immediately rewarding. The child improved steadily under the therapy and began enjoying the most normal health of her eight years of life.

Two years of testing proved that Merck's new cholesterol fighter was capable of achieving remarkable results, and the United States Approved Names Council decided it deserved a generic name. What its discoverers had called mevinolin would henceforth be known, generically, as *lovastatin*. When it won FDA approval, Merck would market it under the brand name *Mevacor*.

Dr. Tobert and his colleagues had made several trips to Washington to hold seminars and give various presentations to keep the FDA advised on the details of the lovastatin testing. Merck's innovators had compiled detailed information on more than 1,200 patients treated with lovastatin in addition to the long-term safety studies in animals. More than 160 volumes of printed documentation on the safety and effectiveness of lovastatin were assembled, enough to fill "a small truck," and a formal New Drug Application was submitted to the FDA in November 1986. Some weeks later the FDA gave lovastatin its formal "1A" classification, meaning it would be judged as a high-priority, possibly breakthrough drug.

Merck's scientists were asked to appear the following February before an FDA advisory panel, a group of independent experts who review all data on experimental drugs and make recommendations to the FDA regarding therapeutic value, safety, and efficacy. All those scheduled to appear prepared carefully for the meeting, fully aware that advisory committees can sometimes be unpredictable. Years of effort and large sums of money could go down the drain if the advisory panel voted against approval, because the FDA usually goes along with the panel's recommendation.

A contingent of Merck experts joined Dr. Tobert in appearing before the panel, including Dr. Edward Scolnick, president of Merck Sharp & Dohme Laboratories; Dr. Eve Slater, of Biochemical Endocrinology; and Dr. James McDonald, of Safety Assessment. Also available for supporting testimony was a group of outside experts: Drs. Goldstein and Brown, the Nobel laureates from Texas, and Dr. Illingworth, one of the drug's earliest champions, along with others who specialized in eye and liver problems.

The statistical results of the clinical testing were undeniably impressive with reductions of between 19 and 42 percent in low-density lipoprotein demonstrated. The reduction in total cholesterol ranged from 18 to 34 percent with marked reductions in very-low-density lipoprotein cholesterol and triglycerides, other highly damaging blood fats. On the other hand, levels of the so-called good cholesterol, high-density lipoprotein (HDL), that helps rid the body of harmful fats, were shown to rise by as much as 13 percent.

Lovastatin passed the test for negative side effects with high marks. In controlled studies, less than 1 percent of the patients treated were unable to continue the program, indicating that lovastatin was unusually well tolerated. Minor adverse reactions—gastrointestinal discomfort, rash, and headache—proved to be mild and transient. The only other concerns related to the slight possibility of long-term problems in the liver and eyes.

Some 2 percent of the patients treated with Mevacor showed substantial elevation of liver enzymes that proved to be symptomless and reversible. The rumors that compactin had caused tumors in dogs compelled Merck's researchers to conduct extensive toxicity studies on the animals with lovastatin. The tested dogs were given extremely high doses of the agent; no tumors were found, but a few developed cataracts. In testing with human subjects, many of whom had been taking lovastatin for periods of one to four years, no causal relationship between the drug and cataracts was found.

At the end of the long February afternoon, the advisory panel voted unanimously to recommend that the FDA approve Merck's new anticholesterol drug. Formal approval came a bit more than six months later, on August 31, 1987. As a precaution, the FDA ruled that Merck's labeling should advise doctors to conduct regular liver tests and annual eye examinations for all who used the product. It was officially approved for use by patients with dangerous levels of cholesterol that could not be reduced by diet or other non-pharmacological therapy.

In a formal press conference in Washington the following day, Merck & Company announced the news to the world. The importance of this new "magic bullet" against a major destroyer of human life was demonstrated by the extensive TV and newspaper coverage the story received around the world. Innovative planning and production techniques enabled Merck to ship supplies of Mevacor tablets to pharmacists two weeks later, some ten months after it had sent its formal application to the FDA.

THE INNOVATION EQUATION AT MERCK

In annual *Fortune* magazine polls of executives, outside directors, and financial analysts, Merck had repeatedly finished first among companies in the pharmaceutical field and was named the most admired of all American corporations in 1987 and 1988. Many industry observers suggest that the arrival of Dr. P. Roy Vagelos was the real turning point in Merck's search for an effective anticholesterol agent and its rise to the top of the global pharmaceutical industry. Vagelos's impact on the company is reflected in his elevation from head of research in 1975 to the office of chairman, president, and chief executive officer in little more than ten years.

Adding color to the Vagelos story is the fact that this son of immigrant Greek parents grew up in and around Rahway, the relatively small New Jersey town where Merck & Company, Inc., is headquartered. The Merck chemists who visted his parents' Rahway café inspired him to study chemistry at the University of Pennsylvania, but, even after he had earned his medical degree and later accepted a research position at the National Heart Institute, he had no idea that his expertise in the mechanisms of lipid synthesis would ever lead him to Merck.

> When Al Alberts and I first came here, we took a risk and the company took an even bigger one. Our risk was saying there's a very special way— using our kind of science—to discover drugs. Basically, I was brought in to increase the impact of modern biochemistry and enzymology. The company didn't ask me to do this; I simply said I'd like to do what I do best. The top people I talked to, Henry Gadsden and others, knew very little about science and were just betting on me as an individual. It was a big risk for them because I was an *academic* scientist, and it was a risk for Al and me because we knew that we could continue to succeed and be productive in the academic sphere.
>
> My best friend at Washington University warned me that I might end up selling toothbrushes and combs! My reason for taking the job was that it was an enormous challenge, and I am very turned on by challenges. Could this kind of research and knowledge be utilized to actually bring me back to what I had really started out to do—to take care of patients? Could I take that biochemistry and enzymology and make drugs that would affect diseases? Of course, failure would have been bad for me and the company.

His enzymological approach to research enabled Merck's scientists to isolate lovastatin quickly, but only after four years had been spent setting up the new systems. Those basic systems, with updates, continue in use and have produced other important drug discoveries. Vagelos also began to recruit new people to man that program, having first learned the value of good recruitment when he hired Al Alberts as his personal assistant years earlier in Bethesda. The ability to recognize, recruit, and reward special talent is a key component of successful innovation in any field, and Vagelos credits Merck's ability to develop talent as a primary reason for the company's emergence from the economic doldrums of the early 1980s to its preeminent worldwide position today.

> I have this terrific anxiety about innovative things so we are constantly looking to see what is happening in all the laboratories of the world. We would like to capture the latest basic research and apply it to drug discovery. To be able to do that you have to have the very best people who know biochemistry, chemistry, microbiology, molecular biology, physiology, pharmacology, and all the other disciplines. We need to have people who are up-to-date in their fields—who can jump right in and apply any new knowledge as soon as it is announced.

Merck's scientific work force has doubled since Vagelos joined the company, and the primary recruitment chores are handled today by Dr. Scolnick, MSDRL president, whom Vagelos recruited a number of years ago to head up the basic research in the department of virology and cell biology. Keeping the talent pipeline open while assuring the productivity and happiness of veterans and newcomers is an ongoing effort at Merck. Innovative work often yields surprising rewards, including occasional bypasses of the usual promotion sequence and elevation to unexpected heights when merited. Recognizing the modern dilemma of employees with young children, the company allows some to work hours of their own choosing and also provides a day-care center for employee children.

Merck announced record revenues of $5.1 billion in 1987 with profits of $906.4 million, and more than half of its sales were made abroad. The company projects that Mevacor will account for more than a billion dollars in annual sales by 1992. Critics claim the per-tablet price of the drug will cost some patients upward of $3,000 per year, making it too expensive for many who need it most. Merck contends that it cost more than $125 million to bring the drug to the marketplace, and 11 percent of its revenues go back into research and development to find new cures. That reinvestment is larger than that of any other drug firm.

Every pharmaceutical company has an inherent need to produce new and better products in order to remain financially viable, as Merck has demonstrated by its effort to find an even better anticholesterol agent. Chemist Robert Smith turned his attention to the possibilities of new forms of lovastatin back in 1984. He and his colleagues succeeded in creating one they named *Zocor*, a semisynthetic derivative of Mevacor, and animal testing indicated that it was even more potent than Mevacor. Merck patented Zocor and began clinical studies with it in 1985.

Merck had more than one hundred drugs on the market in 1988, ranging from antibiotics to Recombivax HB, a vaccine for the prevention of hepatitis B, which was the first vaccine for humans developed from genetic engineering technology to be licensed anywhere. On average, more than 100,000 compounds have to be tested in order to file some 100 New Drug Applications with the FDA for permission to begin clinical testing. About *1* of the 100 tested compounds ever becomes a commercial product, and the average drug spends about ten years in the R&D stage.

These factors and the shortening of the life of patents, the industry contends, drive drug prices ever upward. Federal regulations in the United States doubled the testing period for new drugs in 1962, reducing the life of a company's patent from seventeen to an average of about seven years by the early 1980s. This allows less innovative competitors to avoid huge R&D costs and manufacture copies that sell for less than the original.

The Office of Health Economics studied the drug industry in the United States, Switzerland, West Germany, the United Kingdom, Japan, France, and Italy and reported, "The past success of the pharmaceutical industry in these countries has been due largely to a good balance between the need to encourage a profitable research-based industry and the need to keep pharmaceutical expenditures down to reasonable limits." That, Merck's admirers claim, is exactly what the company has managed to do.

Merck officials are quick to stress that medicines comprise little more than 5 percent of America's annual expenditure for health care and remind critics of their dedication to the discovery of so-called public service or orphan drugs that are used to treat rare diseases, sometimes allowing the company little opportunity to recoup its investment in research and development. Merck's intention to follow through on the Vagelos ideal of taking "care of patients" was demonstrated by the decision to cooperate with the World Health Organization in supplying one of its major drug discoveries free of charge to people stricken with a devastating tropical disease in some thirty-five developing nations. Ivermectin MSD is being donated by Merck for treatment of oncocerciasis (river blindness), which threatens an estimated 85 million people in Africa, the Middle East, and Central and South America.

Ongoing research seeking treatments for cancer, AIDS, prostate problems, diabetes, arthritis, and other diseases is indicative of Merck & Company's scientific and financial vigor. The corporation's decision to strike out in new directions by investing time and capital in setting up different systems that led to the discovery of lovastatin and other significant drugs lends credence to the theory that innovation can pay huge dividends—even in the biggest organizations.

Dr. William Shockley, left, Dr. Walter M. Brattain, and Dr. John Bardeen, right, inventors of the transistor, are shown at Bell Laboratories in 1948, discussing the crystal structure of semiconductive materials.

(BELL LABORATORIES/COURTESY OF AMERICAN INSTITUTE OF PHYSICS, NIELS BOHR LIBRARY)

DATE Dec 24 1947
CASE No. 3 9/39-7

We attained the following A. C. values at 1 ooo cycles.

$E_g = .016$ R. M. S. volts $E_p = 1.5$ R.M.S volts

$P_g = \dfrac{?}{5.4 \times 10^{-7} \text{ watts}}$ $P_p = 2.25 \times 10^{-5}$

Voltage gain 100 Power gain 40

Current loss $\dfrac{1}{2.5}$

This unit was then connected in the following circuit.

This circuit was actually spoken over. and by switching the twice in and out a distinct gain in speech level could be heard and seen on the scope presentation with no noticeable change in power quality. By measurements at a fixed frequency

A laboratory notebook entry by Walter H. Brattain records the events of December 24, 1947, when the transistor effect was first observed.

Frederick W. Smith, chief executive officer of Federal Express, 1988.

Dr. Jean F. Borel has been called the father of cyclosporine, a drug useful in suppressing the body's rejection of transplanted organs.

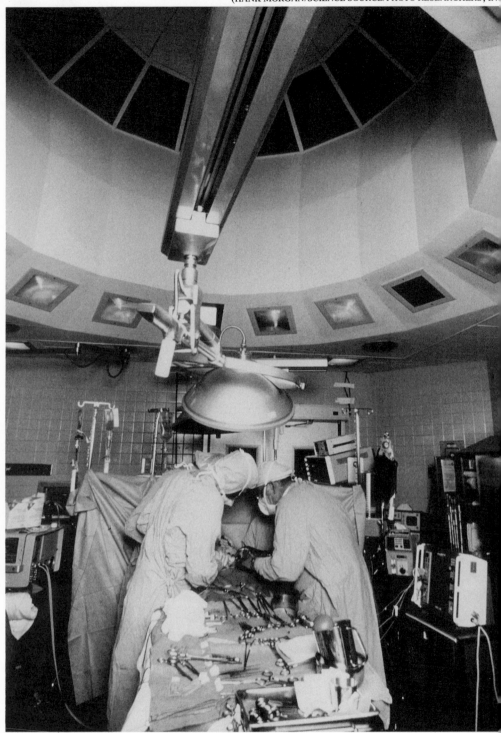

Dr. John Najarian conducts a kidney transplant operation at
the University of Minnesota Hospital.

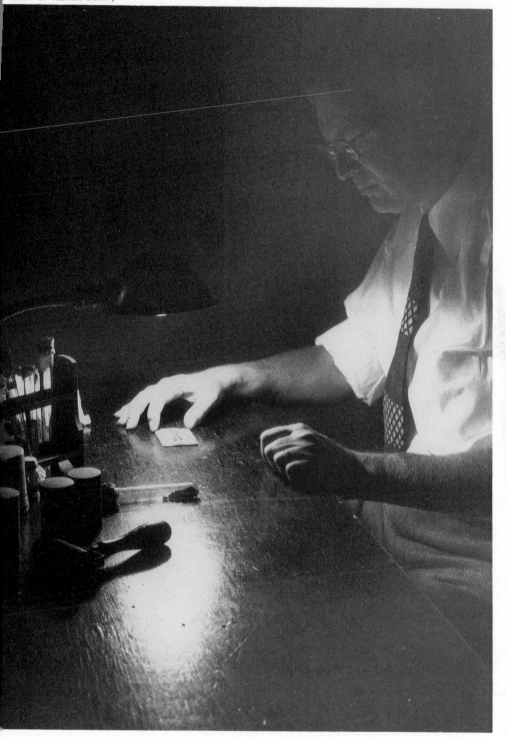

In 1963, Chester F. Carlson re-created the experiment that produced the first xerographic copy by exposing the inscribed slide and charge plate.

Chester Carlson, center, inventor of xerography, is shown in a 1948 photograph demonstrating an early xerographic printer for Dr. John H. Dessauer, left, head of Haloid's research, and Joseph C. Wilson, Haloid's president.

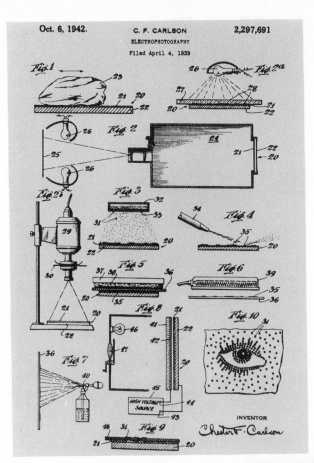

Above: Chester Carlson's original patent, describing his invention of electrophotography, was issued October 6, 1942. *Right*: The 1949 copier Model A was the first commercial xerographic copier machine.

Above: In this photograph from 1960, Dr. Theodore H. Maiman studies the "light wonder" he helped develop at the Hughes Research Laboratories, a cube-shaped ruby crystal that forms the heart of a laser. *Opposite top*: Dr. Charles H. Townes, left, and an associate, Dr. J. P. Gordon, are pictured with the first maser, which they constructed at the Radiation Laboratory of Columbia University, 1958. *Opposite bottom*: This microscopic crater, magnified 480 times, was caused by the impact of a laser.

Dr. Arthur L. Schawlow makes a fanciful demonstration of
laser application in 1972.

A technician adjusts a laser device used to measure precise
radio in a near-field antenna test facility.

Donald B. Keck, left, Robert D. Maurer, center, and Peter C.
Schultz, Corning's fiber-optics innovators, in 1970.

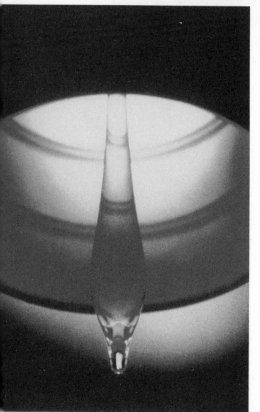

Above: A technician observes an inside vapor-deposition lathe in operation. *Left*: A "gob" falls free from the heated end of a silica blank, beginning the draw process—a critical step in the production of optical fiber.

Citibank's automatic teller machine translates for China-town's non–English speaking residents, New York City.

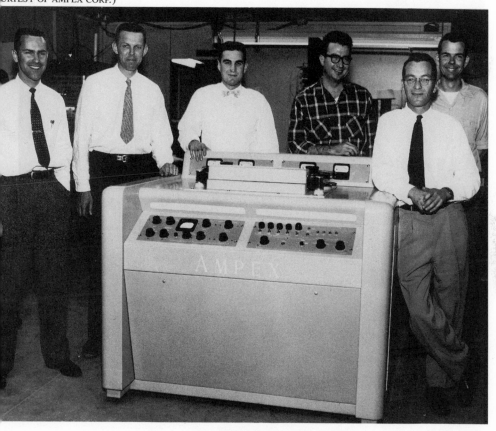

The developers of the first videotape recorder pose next to the Ampex VR-1000 in 1957. The team leader, Charles P. Ginsburg, is second from right.

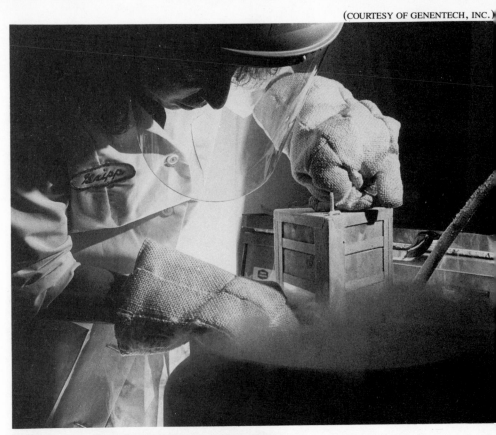

Above: Microbial and tissue culture cells are preserved in liquid nitrogen for later use by Genentech scientists, 1986. *Left*: A Genentech scientist analyzes laboratory assays to help determine the effectiveness of a pharmaceutical produced through recombinant DNA technology, 1986.

Staff members monitor conditions for manufacturing proteins, 1986.

Vials of freeze-dried products produced from recombinant
DNA technology are readied for labeling at Genentech,
1986.

9

The Electronic Bankers: The Rise and Spread of ATMs

"In a business in which convenience is the utmost benefit we can provide for our customers, the use of electronics opens vast new avenues. Prior to ATMs, people were limited to bank hours. Now they have the ability to do their banking twenty-four hours per day, seven days a week."
— JAMES W. JOHNSTON, CITIBANK

Traditionally among the major businesses least susceptible to change, banking first moved toward modern technology in the mid-1950s with the use of a device that permitted the reading of magnetic characters. The resultant acceleration of processing systems enabled banks to encourage the opening of checking accounts by countless millions of Americans who had never before had them. But the great thrust toward electronic banking began with the arrival of the automatic teller machine (ATM), a wonderfully user-friendly device that has permitted customers to do most of their banking at times and places of their own choosing.

Many people and organizations in the United States and abroad have made significant contributions to the development of electronic banking, as demonstrated by the automatic cash dispensers that were first tested in Japan and Europe. Those who remember the earliest Japanese device say it resembled a milk-vending machine and operated with the use of small plastic entry cards that had more in common with poker chips than today's bank cards. The machine captured the card, in exchange for the cash, creating new accounting headaches for the bank.

The cash dispensing machines deployed by banks in England, Austria, and Scandinavia, also in the mid-1960s, enabled depositors to withdraw cash twenty-four hours a day at the risk of considerable losses through fraud. The computerized British machine, built by Thomas De La Rue Ltd., relied on a paper punch card as its means of entry. The customer was obliged to carry the card—without folding, spindling, or mutilating it—and thieves quickly thwarted the meager security precautions. The Japanese and European machines did nothing more than dispense cash, but they attracted considerable attention in the United States. In fact, a revised version of the British cash dispenser was actually tested for a brief time at the First Pennsylvania Bank in Philadelphia in 1968.

With the tremendous increase in checking transactions, predictions that banks would soon be inundated in paper abounded. A number of companies began experimenting with methods of making a machine that might reduce those transactions and save money. Diebold Incorporated (not affiliated with the author) produced a prototype of a cash dispenser at a banker's conference in 1957. A Texas company, Recognition Equipment Inc. (REI), also thought the idea of an automated teller a potentially profitable one and dispatched executive Don Wetzel to Britain to study the De La Rue machine. Wetzel returned to Texas confident that REI had the technical capability to produce a better and far more secure machine.

The task of designing the machine was given to a division of Recognition Equipment that would later become the Docutel Corporation, and Jack Chang was named to head the developmental team. Chang and his colleagues believed the security problem could be solved by the use of a new technology that they had helped develop. They set out to build a solid machine that could dispense cash and perform other tasks when its computerized components were activated by the depositor's plastic entry card, similar to charge cards that banks were beginning to issue. Key to the security issue would be a magnetic stripe attached to the back of the card that would contain intricately scrambled information, including the depositor's personal identification number (PIN), that would allow only him or her to gain access to his or her bank account. Chang would later win patents for the components of the machine.

REI was a client of New York's Chemical Bank, and, when the machine's development was well under way, company executives traveled to New York for consultations with Chemical's board of directors. After much discussion, the bank agreed to pioneer the use of Docutel's new automatic teller machines. The bank's leaders were attracted to the device's convenience factor and its ability to dispense emergency cash to customers but apparently failed to recognize its huge potential as a revolutionary innovation for the long-term future of banking. A young member of the marketing staff, Gordon Piercy, was named to represent the bank's interests in the final development of the first viable ATM. Piercy remembers being handed the job because his supervisor's boss didn't want him to waste any time on it.

CHEMICAL PIONEERS THE ATM

Gordon Piercy spent much of 1969 traveling between New York City and Texas, making certain that Docutel built into its new automatic teller machine all the features that would make it of optimum value to the bank. Development of the machine's mechanics went well, but Piercy recalls that problems with the magnetic stripe were particularly troubling— from finding a method of making the stripe adhere to the plastic card to refining the means of encoding the necessary information. He once put a sample card into his hip pocket and deliberately sat over the electric motor of a New York subway car in a personal experiment to determine whether large electric or magnetic fields might erase the information on the stripe. The subway did not.

The mystery and newness of the magnetic stripe also created controversy when the machine was nearly ready to go into operation. Chemical Bank wanted to work out an arrangement with the Mastercharge people, who had been issuing credit cards to the bank's best customers for only a year or so, that would allow the bank to put its magnetic stripe on the back of the charge card so it could double as an entry card for its ATM. The Mastercharge security chief questioned the safety of the new technology and initially opposed the idea. However, Docutel created a mechanical device, the Docucoder, that allowed Mastercharge's own people to apply the codes at its company headquarters in Lake Success, New York.

A lengthy survey was taken to pinpoint the best place to put the Chemical Bank automatic teller machine, and it was decided to bypass crowded Manhattan in favor of a surburban branch. The Chemical branch in Rockville Center, Long Island, served a bustling community of city commuters who seemed likely to be most appreciative of a machine that could provide them with instant cash twenty-four hours a day. America's first magnetic-stripe ATM was officially installed in the Rockville Center branch of Chemical Bank in September 1969.

Despite the ready acceptance of the machine by the public, especially on weekends and after hours, the first Docutel ATM was a relatively primitive machine by today's standards. A Docutel engineer took up residence in Rockville Center to keep it operating smoothly and made frequent on-the-spot design changes that were sent up from Texas during the early weeks of the program. In the beginning, customers could make two daily withdrawals of either $15 or $30. The totals were later raised to $25 and $50, but the automatic dispensing of the money was an ongoing problem. To prevent the machine from malfunctioning, only brand-new bills from the Federal Reserve Bank could be used and they had to be placed into envelopes and hand-fed periodically into one of two canisters inside the ATM. The first machine had no vacuum-lifting device but depended upon a mechanical arm and wheel to push the designated envelope from the bottom of the big stack into the outer drawer where the customer could retrieve it. The weight of the money-laden envelopes on top often caused the machine to jam, and the Fed found it increasingly difficult to supply mint-fresh bills when ATMs became more popular. Except for the confusion about the cash coming from their checking rather than their charge accounts, customers seemed to have little difficulty with the two-button operation of the ATM and became almost addicted to its simplicity after their first use.

Chemical Bank viewed its initial experience with automatic tellers as a qualified success, and moved to expand its program. Docutel, on the other hand, profited greatly from its Chemical Bank field testing. Having heard about the new banking technology, a steady stream of representatives from banks across the country began paying visits to Rockville Center to see the new automated wonder in action. What they saw was a clear-cut method of reducing human teller work and costs and supplying retail customers with a marvelous convenience. At the same time Docutel's innovators had uncovered a rich vein to mine that caused them to split off from parent REI. Under Jack Meredith's leadership, the new corporation became the leader of a growth industry. Within three years, Docutel had five hundred of its ATMs operating in some 120 banks across the nation.

A few banks, particularly in Washington state and Massachusetts, also saw the ATMs' potential and deployed large numbers of them quickly. Chemical Bank's pioneering ATM program grew more slowly during the early 1970s, but it remained well ahead of most of the others. One reason for Chemical's relative slowness in expanding its ATM network throughout New York was the discovery of an unexpected security problem—one born of human miscalculation rather than technological inadequacy. The strategy of the data-laden magnetic stripe provided the security needed, but the decision to issue the customer a PIN number that had no meaning to him—it wasn't his birthday, address, or what-have-you—meant that few depositors took the trouble to memorize it. Many wrote it either on the back of their entry card or on a slip of paper they kept with it. When those items were lost or stolen, thieves soon recognized that they had direct access to the victim's checking account.

The Docuteller machine had inside it a "hot card" list capacity where information could be filed on up to three hundred missing cards. When one of the suspect cards was used, the machine would recognize and capture it. But that limited list had to be continually updated by laboriously inserting the missing numbers one at a time by hand in each individual ATM in the network, and it proved to be grossly incapable of keeping pace with New York's criminal element. Fraud losses mounted in direct proportion to the increase in machines and cardholders. The bank was obligated to reimburse the customer for everything but the first $50 lost through a missing card, but Chemical spokesmen say they don't remember an instance where the entire loss was not refunded. Into the early 1970s the bank's losses through missing entry cards were so large that the figures remain secret.

Chemical Bank had been unable to put its growing network of ATMs on line with a mainframe computer up to that point, but increasing fraud losses and the expense of operations to correct the problem through the individual ATMs made it imperative. When the ATMs finally went on line in 1974, the central computer's ability to provide instant updates and other services solved many problems. Fraud losses virtually ended—for a time. The bank was attempting to conduct all of its business through two mainframes, meaning that the ATMs were taken off line at regular intervals, and work crews often severed the connecting lines when they dug up the city's streets. Those and other disruptions provided people with larcenous intent a new opportunity to penetrate Chemical's defenses. Knowledgeable thieves learned to insert an entry card into the ATM and first request a balance check. If the machine could not provide an instant answer, they knew it was not connected to the mainframe, and they were free to make unreported withdrawals without hindrance. An excellent feature of Docutel's stripe three technology was its ability to place limits on transactions—two per day for a maximum of $200—but fraud losses began another steady upward climb. The bank's management was forced to consider whether or not the convenience of the $30,000 machines was being negated by the serious problems they seemed to induce.

Chemical Bank, America's ATM pioneer, had 39 machines deployed by 1975, compared to the national average of only 20 per institution. By the end of that year, a total of 4,000 ATMs had been installed in banks across the United States. Chemical's new generation of Docutel machines were offering a variety of services in addition to cash dispensing, but the New York bank decided to put its ATM network on hold.

THE CITIBANK APPROACH

Citibank, in the mid-1970s, was clearly one of the strongest commercial banks in New York City and the nation, but it had not yet reached the dominant position it occupies today. It had paid close attention to the ATM programs of rival Chemical Bank and others but had deliberately not rushed into major deployment of its own. That attitude changed, seemingly overnight, in 1977.

Walter Wriston, who was Citibank's chairman during its period of greatest growth, has received universal kudos for his stewardship. One analyst, who prefers to remain anonymous, told us:

Wriston was something of a cowboy when he was running Citibank. My version of his theory of operation is that he set up his organization like Chairman Mao might have done it—"Let a thousand flowers grow"—and he made it immensely competitive internally. So the people at Citibank would not only be competing with rival banks but also with other units inside their own organization. He created an environment, as it was often explained to me, where you were allowed to succeed and allowed to fail. What you were *not* allowed to do was to do *nothing*.

Many also explain Wriston's successful tenure by pointing to his uncanny talent for picking the right people to do the necessary jobs. Most agree that Wriston's selection of a man named John Reed was another prime factor in Citibank's ability to make such deep inroads into consumer banking, borne out by Reed's elevation to the Citibank chairman's office on Wriston's retirement. As head of the Citibank operating group, Reed told Wriston that he was concerned that the bank's consumer business had no separate or specific plan that would permit it to be developed as a revenue stream. A new approach might allow it to produce a quarter or more of company net income in a few years.

Reed's concept held that the bank could do considerably better by reducing its unit costs. If the unit cost could be sufficiently lowered, it would be possible to increase revenues by attracting many more low-balance customers. The key to financial growth, from Reed's viewpoint, was hidden in the bank's consumer business. As elementary as this theory may seem, it was new and basically untested in banking, and Wriston gave Reed permission to explore it. Reed's first move was to bring in two experts in the package marketing field, Rick Braddock and Pei Chia.

Braddock and Chia began a number of complex studies that enabled them to predict the most advantageous methods of increasing Citibank's consumer business. Despite the ATM's then-status as a stepchild in the banking world, Chia came to believe that ATMs could perform a number of transactions that might attract the kind of hurried, harried New York consumer Reed hoped to reach. Wriston and Reed wanted visible proof of Chia's theory.

Tom Metz, now a vice president of Electronic Data Systems, had started with Docutel in 1970 and actually installed the seventh automatic teller machine sold in the United States. He has vivid recollections of the thoroughness of Citibank's ATM research.

Citibank bought its machines from Docutel in the very beginning, and, since I was the account manager on that job, I was asked to teach them everything I knew about the concept and potential of ATMs. Reed seemed to be a big believer in ATMs, although there were also a lot of skeptics at Citibank at the time. To really test the acceptability of the program they had in mind, under Reed's direction the bank contracted for a local company to actually build a mock branch in the basement of the Daily News Building. For about a year, as I remember, they invited customers in to do a lot of their standard transactions through automatic teller machines. And they constantly changed the configuration of the branch, trying out different machines, signs, lines, etc. All the time they kept watching the customers to get their reactions—how they entered the branch, what colors they preferred, which buttons they most often pushed, which signs got the most attention—and asked questions about the service. They really spent a lot of dollars in order to come up with the manner in which to best support this kind of automated branching concept. When they finally deployed their machines, they had a pretty comfortable feeling about how the customers would support the system.

That deployment, after many months of research and development on both the marketing and the engineering side, was on a scale that startled the banking industry. Citibank's research indicated that a new kind of automatic teller machine was needed, a machine that had to be designed as an extension of the teller's counter—an ATM that could provide full teller service. Because no such machine was then available, Citicorp designed its own and called it a customer-activated terminal (CAT), subcontracted the development of component parts to outside companies, and assigned the assembly of the machine to its own California subsidiary, Transaction Technologies.

Interestingly, one of the companies that produced components for Citibank's ATM was Ohio's Diebold Incorporated. Although it had created an early ATM prototype, Diebold lagged far behind Docutel and others in manufacture and sales in the mid-1970s, and Citibank's large order helped propel it on an upward spiral to the point where it would control about 47 percent of the ATM market in the United States by 1988. Docutel, whose innovators had created the magnetic stripe technology that became standard in ATMs, fell by the wayside. After a history of leading the industry for more than ten years, Docutel merged with Italy's Olivetti, was eventually sold to TRW, and now no longer makes ATMs in the United States.

Citibank launched its ATM-oriented consumer-banking offensive on an unprecedented level. No official figures about the cost of the bank's original CAT-1 deployment have been released, but most banking analysts believe a minimum of $250 million was invested. Instead of installing a few score ATMs in its branches, Citibank deployed approximately five hundred of the most versatile ATMs yet seen: at least two in every Citibank branch in the metropolitan area. Cognizant of the well-founded fears of security-minded New Yorkers, Citibank stationed most of its ATMs in special vestibules where customers could make their transactions behind the safety of locked doors at any hour of the day or night. With that number of ATMs available in safe, convenient locations all over New York and its suburbs, electronic banking suddenly became an extremely attractive option for many who had never before used Citibank's services. To alert the public about the advantages of its new automated services, Citibank launched a major television and radio advertising campaign that stressed the memorable slogan "THE CITI NEVER SLEEPS."

Much of the banking community, not just in New York but all across the nation, was convinced that Citibank had gone overboard in its dedication to automatic teller machines. The consensus was that ATMs were attractive conveniences for the customer, but only brick-and-mortar—construction of new branches—was capable of actually increasing market share. Citibank's vast CAT-1 program quickly disproved that theory. The bank's share of retail deposits was a meager 4.7 percent when the ATM program was inaugurated in the summer of 1977, but that share had risen dramatically by 1979 and gave every indication that it would continue to escalate. Competing New York bankers, including the ATM trailblazers at Chemical Bank, were forced to take another look at their own automatic teller programs.

NETWORKS

Phil Megna, now a Chemical Bank vice president, was assigned the task of studying the deficiencies of Chemical's dormant ATM program and devising methods by which the bank might be able to make up market share lost to Citibank. Megna took on the job in 1980 and immediately instituted extensive studies that examined every aspect of Chemical's existing program with its thirty-nine ATMs. In less than two years, all of the bank's Docutellers were retrofitted to bring them up to current standards; advanced machines from a variety of sources were tested; detailed location maps were drawn up to pinpoint the optimum locations for automatic tellers; and more multifunction ATMs were deployed.

When interviewed for this book, Megna said:

We made a lot of mistakes in the early days and I think those who came later benefited by the knowledge gained from our problems. Of course, when you're the first to do something, "gaining experience" might be a more appropriate term than "making mistakes." But we did a lot of things that weren't correct—certainly those machines should have gone on line just as soon as the capability to do it was there. By 1982 we thought enough of our

ATM network to put it on a separate, dedicated TANDEM computer system. Since then, our on-line time exceeds 99 percent, and our fraud losses have essentially disappeared.

Putting its ATM network under the control of the advanced TANDEM mainframe solved many of the problems that had plagued Chemical Bank since 1969. The gradual expansion of the bank's ATM program and a subsequent advertising campaign promoting the convenience of automated transactions at growing numbers of locations did much toward restoring Chemical's consumer banking interests to better health. However, Megna and his associates had no easy solution to the principal problem he had been selected to solve. The revitalized program was looking good, but if Chemical Bank wanted to be a major player in the ATM game, to compete on equal terms with its chief rivals, it would be forced to invest many millions of dollars in the purchase and proper installation of hundreds of automatic teller machines and millions more in annual operating costs. That is exactly what Megna told senior management when he delivered his official report on May 24, 1982. As he had expected, the bank was not prepared to make an immediate investment of that magnitude. Megna was told to come up with a more manageable plan.

Liam Carmody, president and CEO of Carmody & Co., was brought in as a consultant, and he and Megna began examining an idea that did fit management's instructions. One way that Chemical and other area banks could compete with Citibank—which had more than doubled its share of retail deposits by this time—was to organize a co-operative venture, a network. By sharing the start-up costs and their ATMs, member banks could negate Citi's huge advantage at a much more reasonable price. Megna brought the new plan to management and this time won its support.

Megna and Carmody knew an attempt to start a New York banking network would be difficult, but they both believed it could be done. Such networks had been successful in other parts of the country, and Carmody had been intimately involved in such work since 1974 when he helped set up the first one, American Express' traveler's check ATM network, at major airports across the United States and around the world. Their biggest challenge would be fostering a climate of trust that might overcome some of the intense competition that was traditional among New York banks. The only previous attempt to organize a somewhat similar cooperative venture in the New York area, a point-of-sale (POS) network, had failed.

Banks in the metropolitan area with locations and other assets that would lend strength to the network were targeted as potential sharing partners, and concepts that would allow them to participate in several ways were created. When the detailed blueprint of the New York Cash Exchange (NYCE) was completed, Megna and Carmody began a marathon effort to sell the selected banks on the proposal. They were pleased to encounter less resistance than they had expected. Virtually every bank they contacted expressed genuine interest, but, as time went on, they noticed an inexplicable reluctance of two key banks to join in and permit operations to begin. The indecision was caused by some unexpected competition. Manufacturers Hanover Trust had entered into a joint venture with ATM vendor NCR to form another New York network, creating understandable confusion in the minds of prospective members about which network to join.

"It became patently clear," Megna told us, "that we needed to take a walk over to Manufacturers Hanover and discuss the situation. They weren't getting off the ground because people were wondering if we might have the better system, and we weren't able to get going because others were worried that theirs might be better. So that's what we did, suggesting that we would be better off doing this thing together. They agreed and that was the start of the NYCE network."

Agreement was reached in autumn 1983 to begin the preparatory work for NYCE, and Carmody's company was hired to supervise the entire operation. Liam Carmody's work so impressed the sharing partners that he was asked to serve as the first president of the network when they couldn't agree on one of their own to fill the office. The demands of his own business prompted Carmody to step down two years later, and a member of his firm was named to replace him. NYCE began operations in March 1985 with eight sharing partners: Chemical Bank, Manufacturers Hanover, National Westminster Bank, Bank of New York, Goldome, Marine Midland, Barclay's Bank of New York, and Union Trust Bank of Connecticut. Citibank declined an invitation to join, but Chase Manhattan became the ninth equity partner a year later, sharing network profits with the eight founding members.

Despite its late start, NYCE became the largest shared ATM network in the United States in just three years and trails only France's Carte Bleue in world ranking. By mid-1988, NYCE had nearly 11 million cardholders, who had access to about 6,000 ATMs owned by 360 banks in 22 states from New York to California plus Puerto Rico and the Virgin Islands. At that time NYCE was handling about 6.5 million switched transactions per month. Illustrative of the incredible technology that makes electronic banking possible is the location of the NYCE *switch* or electronic processor—in Brandier, Wisconsin, nearly a thousand miles from the area banks that organized the network. Carmody explains the way the entire process works in eloquently simple terms.

> Let's say an individual who has a Chemical card goes to a machine owned by Manufacturers Hanover. The Hanover ATM recognizes that the card is not one of its own but has been set to find out if it is a valid NYCE card. The Hanover ATM sends the information on to the NYCE switch in Wisconsin, which acknowledges the requested withdrawal as a valid NYCE transaction through MHT on Chemical Bank. The Wisconsin switch sends the request to Chemical's central computer in New Jersey, which checks the customer's balance, debits his account, and sends authorization back to the NYCE switch. The NYCE switch then relays that information back to the Manufacturers Hanover computer, which passes it on to the specific ATM, which delivers the cardholder the money he has requested—almost instantaneously. The entire process takes about five seconds.

This same kind of instant service is available for NYCE cardholders anywhere in the United States where member banks have automatic teller machines. The NYCE network has enabled Chemical Bank to reestablish itself as an ATM innovator. The bank's 39 machines were each averaging about 1,200 transactions per month in 1980. Chemical has since deployed approximately 300 multi-function Omron ATMs from Japan, some of which are averaging more than 30,000 transactions per month. The other New York area network partners have also done well and have launched an aggressive media assault on rival Citibank with commericals that display a piggy bank and proclaim, "IF YOUR BANK ISN'T NYCE, THIS MAY BE YOUR NEAREST CASH MACHINE."

Despite the concerted effort to cut Citibank's lead in retail banking, Citibank's market share has continued to rise by about 1 percent each year since NYCE began operations, as has Chemical's. Since Citi launched its CAT-1 program in 1977 its portion of retail deposits has more than tripled, from 4.7 to 14.7 percent. Citibank was the first U.S. institution to make an effort to deploy hundreds of ATMs, but it has since lost its national lead to two California banks, First Interstate Bancorp and Bank of America, which each had more than 1,350 compared to Citi's 1,281. But that third-place position was destined to be short-lived. In 1988 Citibank began deployment of its CAT-2 program, which called for 950 new ATMs of its own design to be installed throughout the New York area.

James W. Johnston became head of Citicorp's development division in the U.S. consumer services group in December 1984, shortly before the NYCE network was scheduled to begin operations. When interviewed for this book, he told us, "The decision to launch our CAT-2 program is testimony to how well ATMs have worked for us. Most of the original 950 CAT-2 machines we expected to deploy are now in place and we've decided to go beyond that. As an example, in one branch we went from four ATMs to ten; now we're going back to take it from ten to twenty."

Citibank's large-scale advertising campaign refers to its newest ATMs as "success machines," and the public has been advised that they can perform fifty-five separate functions in three different languages: English, Spanish, and

Chinese. The new machines operate on a touch-screen principle, and, as Johnston says, their popularity has caused the bank to multiply their installation in heavy-traffic branches, thus preventing the long lines of the days of the human teller. Industry statistics say ATMs save the average bank less than a penny per transaction in comparison to those handled by human tellers, but Johnston claims Citibank's automated transactions are only about a third as expensive as the routine kinds.

Chemical Bank demonstrated its renewed faith in electronics by opening its branch of the future in suburban White Plains, New York, in April 1988. Experts say it is the closest version to date of a fully automated bank branch with only a head teller on the premises. The new automated branch, on the site of a traditional bank, occupies only about half the space of the latter and provides all the normal teller services. All teller functions are handled by multipurpose ATMs, and alongside the ATMs are customer-activated terminals that perform like personal computers. By walking through a simple menu and touching the screen the customer can obtain information about accounts, rates, loan services, and mortgages, and the pressing of additional buttons produces statements and other related printed material. On-premise telephones give the consumer instant access to a central customer-service number and immediate solutions to problems the machines might not be able to solve. Another Chemical innovation is what it calls corporate cashiering. The bank has set up ATM facilities at fourteen corporate locations of IBM, and other companies, where, for a fee, it provides instant cash and other services normally available only at a bank branch.

Although many industry analysts say the overwhelming majority of U.S. banks are still not taking full advantage of the many benefits of their automatic teller machines, the sheer proliferation of the ubiquitous devices echoes their popularity with the banking public, especially those under the age of thirty-five. The ATM's success is particularly apparent in New York City, thanks to the huge deployment by Citibank and the members of NYCE. Industry surveys reveal that a full 60 percent of those under thirty-five in the New York area use ATMs regularly and estimates are that upward of 75 percent of the nation's bank customers will be steady ATM devotees by the mid-1990s.

BANKING'S ELECTRONIC FUTURE

During the past two decades or so, banks have become second only to the federal government in the purchase and utilization of computers. Electronic devices have brought sweeping changes in traditional banking since the introduction of the automatic teller machine in 1969 and the automated clearinghouse in the mid-1970s. Technology that could soon provide even more dramatic services for the public is already available.

The more innovative banks began testing a form of video banking in the early 1980s in an attempt to transport the new services they were offering directly into the customer's home or office. Citibank's Direct Access service is reportedly the most successful such program in the nation, but Jim Johnston agrees that it has not developed as rapidly as expected. With the coming of mergers between brokerage houses and/or credit card companies and giant retailers with sophisticated national electronics networks, banks faced a new problem: the loss of savings account funds to money management accounts. Several major banks began to counter the threat with management programs of their own. In 1983 Citibank, for example, introduced its Focus account, which combines a money market checking account, a brokerage account, a cash and a tax-exempt money market fund, a preferred Visa card, and an integrated statement. Johnston says both Direct Access and Focus have continued to grow, but neither Citibank nor the other New York institutions that have such programs are actively promoting them. Growth has been slow, largely because the proliferation and home use of personal computers have not met expectations.

One area in which Johnston and virtually every other consumer banking expert sees enormous potential growth is in point-of-sale (POS) networks. ATMs have had relatively little effect on reducing check writing, but POS networks should decrease the paper flow from retail outlets. Better and less expensive technology has already created an explosion of POS terminals on the credit card side, and that basic technology can be used for debit card terminals. Credit card POS terminals jumped from a virtual handful to around 300,000 nationwide between 1980 and 1985, and completely transformed the credit authorization process to the delight of both patrons and retailers. Debit card POS networks would operate on the same general principles, with the fully on-line ATM entry cards replacing the credit card. With the debit card POS terminals in place in retail outlets, a depositor could insert his bank card and have the cost of his purchase automatically deducted from his account, dispensing with the need to carry either a checkbook or cash. But banks have been slow to take advantage of this technology, which could have a more profound effect on the industry than even the automatic teller machine.

As of early 1989, approximately 40,000 to 50,000 debit card POS terminals were in operation nationwide. Most are in California, Arizona, and Florida. New York and other eastern population centers are lagging far behind.

Citibank, so advanced in its ATM program, concedes that POS terminals are the wave of the future, but its cardholders could make purchases in 1988 only at Mobil and Exxon gasoline stations and at Pathmark Supermarkets. Citi's POS program was accounting for several thousand transactions per month at that time with strong indications that the totals would continue to grow. Supermarkets, cash- or check-based stores that don't accept credit cards, have been the favorite testing ground for POS debit terminals. The speed and convenience they provide at the checkout counter have been a boon for both customers and store managers. Ironically, the fact that retailers are so enamored of the new service partially explains why more POS terminals are not available.

Industry analysts say the delay is being caused by an economic rather than a technological issue. The banking and retail industries are still attempting to determine which stands to gain more from the complicated technology and, therefore, which of the two should bear the lion's share of the costs. Once these contentious questions are answered and a general deployment of debit POS terminals begins, the impact on the general public will be enormous.

In the meantime, other organizations are finding uses for multipurpose ATMs and similar machines. A whole range of services are in the planning or developmental stage. The U.S. Postal Service began using ATMs in 1988 to speed about 68 percent of customer business usually handled by postal clerks. In the spring of 1987 Ramsey County, Minnesota, inaugurated a program that allowed recipients of Aid to Families with Dependent Children to receive their monthly payments through ATMs at banks and stores. Some of the machines in the program were equipped with special screens that used symbols to help those unable to read. And a number of retail chains—including Sears, 7-Eleven, Safeway, and Publix Supermarkets—have built their own ATM networks and opened them up to banks on a fee basis.

Because ATMs were developed originally as machines that could provide bank depositors with instant cash, it is somewhat ironic that an executive for the leading American manufacturer of ATMs has publicly proclaimed that similar devices may be at the vanguard of the much-discussed "cashless society." Robert Barone, senior vice president of sales and marketing for Ohio's Diebold Incorporated, believes that new machines will vend all manner of small items, such as books, videocassettes, jewelry, and cosmetics, reducing losses from shoplifting and allowing consumers to make purchases on an around-the-clock basis with the use of bank cards and credit cards.

Consultant Liam Carmody, with a lifetime of experience in all phases of electronic fund transfers, believes current technology is capable of making cash obsolete. In his view, it is a possibility worth close study before point-of-sale networks become fully operational across the United States. Carmody is convinced that the virtual abolition of cash would provide society as a whole many immediate, unexpected benefits that would be of far more importance than the obvious convenience it would offer banks, retailers, and consumers. Carmody points out that

> Hard currency or cash represents a fraction of 1 percent of *money* as it is defined by the Fed. I don't have exact figures, but I would estimate that cash handling represents about half the cost of payments in the banking system. Currency must be kept in huge vaults that are among a bank's major expenditures, and tellers are required to count it, re-count it, and count it again—and nine times out of ten it's cash that causes tellers and

branches to be out of proof. It's easily counter-feitable, and it's expensive for the federal government to produce. If you really think about it, the primary reason we have cash is for a small number of the end points of our retail sales in the United States.

Many people cling to the historic view that cash is an integral part of our lives, but Carmody insists that most cash today is spent for only the smallest items: newspapers, candy, chewing gum, and subway or bus rides. Few people have the desire or ability to pay cash for major purchases such as houses, automobiles, or large appliances; rent, utilities, and most other regular bills are invariably paid by check. Many use credit cards to pay for other expenditures over a few dollars—travel, entertainment, clothing, and the like—so that the move toward a cashless society is well under way. Carmody believes we now have developed the technology to automate small retail outlets, and he thinks the time is fast approaching when we must ask ourselves, "Why do we need cash?" He sees no need for the government to continue its costly printing of large-denomination bills and even questions the real need for bills as large as $100.

One of the reasonable arguments raised against the abandonment of traditional currency is the fear that electronic fund transfers will infringe upon the individual's right to privacy. Such transactions do leave an audit trail, but Carmody feels they would pose a real danger only to those who are working against the general welfare of society. The government, corporations, organizations, and honest individuals are seldom if ever involved in huge cash transactions. That remains the favored method of exchange between tax dodgers and underworld elements. Those who might be disadvantaged by the absence of cash are the poor, to whom cash is the payment of necessity rather than choice. Carmody told us,

> The federal government has made the automation of benefits distribution a national priority. And there's a new concept called Lifeline Banking in which banks can open up fully electronic accounts for lower-income individuals with preauthorized payments for rent and utilities and provide machines that can dispense needed cash. If we can close that loop and take it a step further and make POS terminals available, it would be possible to automate the whole food-stamp distribution system. Proper benefits could be provided for probably half of today's costs at a level of service that would be infinitely better for the people receiving these benefits. Then, if we can overcome the social privacy issue and actually quit printing large currency, I believe we'd have more people paying their fair share of taxes, the underground economy might well disappear, and the entire organized crime and drug culture would have a very severe problem.

Whether one agrees or disagrees, the innovators who created the various systems that made electronic fund transfers possible have opened up an endless array of exciting possibilities—and clearly improved efficiency and convenience for financial institutions and their clients. Nearly all the experts agree that few banks have taken full advantage of the many advances in electronic technology. But a better understanding of the potential, coupled with the falling costs of electronic hardware, seems to indicate that the day is approaching when American ATM cardholders will be able to do virtually everything, from riding the subway to buying groceries, without ever checking the money in their pockets.

10

Audiotape, Videotape, and Videocassette Recorders: American Gifts to the Japanese

"American businessmen try to cash in immediately. They have to make money. But we know that VTR will be important in the future and we are willing to invest now for that time."

—AKIO MORITA, SONY

By the middle of 1988, according to a Gallup survey conducted for the Electronic Industries Association, nearly two-thirds of the households in the United States owned at least one videocassette recorder (VCR) and countless millions more had been sold in countries around the world. A group of innovative California engineers created the first videotape recorder (VTR) in 1956, and a study of videotape technology from that first VTR to today's VCR provides an uncommon example of a variety of approaches in the management of innovation.

Because video recording technology was born in the United States, it is not surprising that appliance stores are overflowing with videocassette recorders bearing familiar American brand names such as Emerson, RCA, Zenith, and General Electric. But closer examination reveals that the mechanism behind the American label is a product of Japanese manufacturing. Beginning with Ampex in 1956, a number of powerful U.S. corporations made major contributions to video recording technology, yet all have dropped out of the multibillion-dollar consumer video market.

The familiar explanation that the Japanese "take our ideas, use cheap labor to reproduce them, and manage to undercut us" offers little comfort in the study of videotape recording because the Japanese added many innovations of their own in design, manufacturing, and marketing. And it must be pointed out that a key facet of the basic technology, the viability of magnetic tape as a recording medium, was first developed in Nazi Germany in the 1930s. An American GI named John T. Mullin "liberated" a German audiotape recorder at the end of World War II and cleared the way for the United States to take the lead in tape-recording technology.

With a civilian background in electronics, Mullin served in the U.S. Army Signal Corps during the war and was posted to Germany at its conclusion. On the advice of a British officer, he went to a Frankfurt radio station to examine a strange recording device and was astounded by what he saw—and heard. The Germans called the machine a *Magnetophon*, and it reproduced sound from a mysterious magnetic tape that was veritably indistinguishable from a live performance. The Allies had nothing like it, and, on the orders of his superiors, Mullin confiscated two of the machines for the Signal Corps. He carefully photographed all the schematics and work manuals for the army and kept extra copies for his own use. He also dismantled two more magnetophons and mailed the pieces home as legal war souvenirs in thirty-five separate packages.

Back in civilian life, Mullin spent four months reassembling and studying the magnetophons and became convinced that the German engineers who had produced them were at least ten years ahead of their American counterparts, who were still experimenting with steel tape and wire. At a formal demonstration of the magnetophon before a meeting of the Institute of Radio Engineers in San Francisco on May 16, 1946, Mullin and his recorder received a standing ovation from the overflow crowd, whose trained ears were unable to detect a difference between the taped music and that of a live group hired to play.

Harold Lindsay, an electronics engineer at the Dalmo Victor Company in nearby San Carlos, was in the audience that night and told Mullin that he would like nothing better than an opportunity to work with magnetic recording someday. His wish was on its way to fulfillment a little more than six months later after an unexpected meeting with Alexander M. Poniatoff, the founder of Ampex. That company supplied the precision magnet motors and generators Lindsay used in assembling the airborne radar Dalmo produced for Sperry Gyroscope and the navy. Poniatoff asked Lindsay to become a part-time consultant to help him find new products Ampex might manufacture at the expiration of its government contracts. He was particularly interested in producing broadcast studio turntables. Lindsay accepted the offer and, after a few weeks on the job, suggested that a magnetic tape recorder might be a better product for broadcasters and aroused Poniatoff's interest by describing Mullin's magnetophon. Once Poniatoff had heard the magnetophon, he agreed with that assessment and asked Lindsay to head the project to develop America's first audiotape recorder.

The tape recorder that Lindsay and associate Myron Stolaroff set out to create was based on the German magnetophon but could not be an exact copy because Jack Mullin never permitted them to see the electronics of his machine. He had personally redesigned much of that system and was negotiating a deal with another company to develop his version of a magnetic tape recorder. Mullin felt the building of a proper playback head would be their most difficult task and agreed that they could make it compatible with the magnetophon, the only machine on which it could be tested. By spring 1947, Lindsay and Stolaroff had built their playback heads, and a formal test on the magnetophon proved they were actually better than the originals.

Ampex's innovators quickly developed record and erase heads and moved on to assembling the rest of the recorder that had been dubbed the Ampex 200. Mullin's plan to produce his own recorder fell through, but his salesmanship was instrumental in creating a ready-made market for the Ampex machine. Bing Crosby, weary of doing live weekly radio shows, had moved to a new network on the understanding that his future programs would be prerecorded by Mullin on magnetic tape. With no machine of his own as a backup to the magnetophon, Mullin asked the singer's people to consider the new Ampex recorder for that job, as soon as it was ready.

Mullin had been permitting Lindsay to use his captured German tape—the only existing magnetic tape available anywhere—but that tape was worn and spliced and the Ampex engineers worked on with the dread question always in the back of their minds: Of what use is a tape recorder without magnetic tape on which to record? The question was answered in dramatic fashion during a short period in midsummer 1947. A representative of a New York firm called Audio Devices asked for permission to test a new magnetic tape his company was developing on the still-un-publicized Ampex recorder. The company eagerly agreed but faced a quandary only a few weeks later when a representative of the 3M Corporation arrived with the same request. Ampex accepted the 3M suggestion that all parties could profit from a cooperative arrangement. Audio Devices later became the eastern distributor of the Ampex 200, and the innovative 3M Company had found itself another new and profitable product that would set industry standards.

Less than a week before the scheduled September demonstration for Bing Crosby Enterprises, the Ampex team began experiencing severe degradation in the record mode and was forced to call Mullin and ask for a postponement. When assured that the playback mode was operating properly, Mullin insisted that the demonstration proceed as scheduled, and Crosby and his people were so impressed with the reproduced sound they didn't even demand a recording demonstration. Representatives of Crosby Enterprises paid a call at Ampex afterward, offered to handle the "West of the Rockies" distribution of the perfected product, and handed Poniatoff a signed order for twenty recorders. A personal Bing Crosby check for $50,000 arrived a few days later, enabling Ampex to start production on America's first audiotape recorders. They were installed at the ABC Studios in New York, Chicago, and Hollywood on April 25, 1948.

THE QUEST FOR A VIDEO COUNTERPART

Professional audiotape recorders elevated the Ampex Corporation to a degree of success that Alexander M. Poniatoff had never envisioned when he founded the company. Poniatoff had used his initials, *AMP*—added the *EX* (for excellence)—and coined the name that had suddenly become synonymous with an important new industry. But new generations of Ampex recorders had barely been installed in radio stations around the world when another broadcast medium, television, began to replace radio as the home entertainment of choice.

The first American television pictures had been transmitted as early as 1927, but TV remained little more than a scientific novelty with no impact on the general public until the late 1940s, when the networks began signing big-name performers and offering better shows that attracted growing audiences. Coaxial cables and other advances enabled TV to spread to wider areas of the United States by the early 1950s, and television was soon facing the same problems that had plagued radio before the arrival of the tape recorder. Programs originating on one coast had to be repeated for showing at a corresponding hour on the other coast, and the only recording method available was the expensive filmed kinescope process, which produced dark, fuzzy pictures.

Television desperately needed a device that would permit the recording of programs for later viewing, and it seemed that the likely solution would be found through the use of magnetic tape. Appropriately, Jack Mullin and electronics engineer Wayne Johnson built the first working prototype of a videotape recorder at Bing Crosby Enterprises in 1951. Mincom, a subsidiary of the 3M Corporation, also demonstrated a video picture that year, and RCA produced another working model in 1954. All three prototypes were little more than audiotape recorders that produced unsatisfactory images when run at incredibly high speeds. Ampex authorized research on a videotape recorder in December 1951 but gave the project a low priority and relatively little funding.

Charles P. Ginsburg was named to head the three-man Ampex team—Charles E. Anderson and Shelby Henderson were the other members—and the management opted to try a rotating-head approach. The engineers were asked to build three heads that would be mounted on the flat surface of a drum, where they would scan the surface of a 2-inch-wide magnetic tape in arcuate fashion. The tape would be set to move at 30 inches per second, producing a head-to-tape speed of close to 2,500 inches per second and, ideally, creating a dependable recording of 2.5-megacycle signals. The work was barely under way when the researchers were taken off the job, in May 1952, to work on another project with a higher priority.

The VTR team produced its first crude recorded picture in October 1952, a little more than two months after the project was allowed to resume. The tenuous success of that first demonstration bought the development team more time and an opportunity to try a different approach, using four heads instead of three, in another prototype that was demonstrated in March 1953. The new model relied on an amplitude-modulation system in which the video signal established the limiting amplitude in a clamp modulator. The picture was far better than the earlier one, but it was flawed by a breakup they labeled "the venetian blind effect" at the points representing the crossover from one head to the next in the reproduced picture. That necessitated extensive revisions to correct the problem, but before those changes could be worked out, the team was again transferred to another project with the understanding they would return to the VTR "within a few months."

Those few months grew to more than a year with little time being allocated to VTR research, except that time Ginsburg and Anderson were able to "bootleg" on their own. Even under these circumstances, the innovators were able to make specific changes in the control system and asked for an opportunity to demonstrate their revamped recorder for a management committee in August 1954. The committee was impressed and authorized a formal resumption of the project on September 1, appropriated more money for the work, and brought in Fred Pfost and Alex Maxey to join the team. The earlier arcuate sweep configuration was dropped in favor of a new approach in which the tape wrapped around the rotating drum, producing information that would be written across the tape in straight lines. To compensate for the amplitude fluctuations characteristic of the rotating-head approach, the engineers began developing an automatic gain control system.

The team had built six heads with ferrite cores and metal tips at the very start of the project, and these heads were still usable. But the high centrifugal force created by the new configuration made them break apart, and Maxey had great difficulty building new ones that would stay together. Despite a number of such problems, the new geometry produced a good picture in December that did not quite meet their expectations. Charles Anderson had been unable to perfect his automatic gain control system, prompting him to suggest that they try dropping the amplitude-modulation system in favor of a vestigial sideband FM system. He started on the FM system the first workday of 1955, and the team was able to demonstrate a very promising first FM picture off tape early in February.

It was at this point that an important contribution was made by a part-time member of the team who had recently returned from military service. Ray Dolby, a college student whose name would later become famous because of his development of noise-reduction systems, had worked on the early VTR program at Ampex on a part-time basis while studying engineering at a nearby college. He became so enthralled with the work that he dropped out of school and immediately lost his draft deferment. Dolby managed to simplify Anderson's FM method by designing and building a multivibrator that could be modulated by applying the composite video signal directly to the control grids. The picture produced with Dolby's multivibrator modulator was even better, and the team felt confident enough to schedule a demonstration of the new prototype for the board of directors on March 2, 1955.

That demonstration so impressed the board that the team was given a new engineering project authorization, much roomier isolated work space, and a specific time frame: a videotape recorder that could be demonstrated publicly in a year's time. Over the next eleven months, Ginsburg, Anderson, Henderson, Pfost, Maxey, and Dolby made remarkable improvements in the basic recorder they had demonstrated in March. The following are just a few of the innovations created during the period: a technique for varying the tape tension solved the problem of information rate changing as the heads wore down to a smaller sweep radius and also helped to make tapes interchangeable from one machine to another; the individual magnetic heads were reconstructed into a sandwich-type design that made them stronger and more reproducible; the bandwidth of the modulation system was extended, enabling it to operate with a carrier system frequency as high as 6 megacycles; the drum was stabilized to allow taped pictures to be shown on a standard monitor; big improvements were made in resolution and signal-to-noise ratio; and the two-mode switching device was replaced by a four-way switcher that allowed conduction from only one channel at a time.

A demonstration before a small group of Ampex executives near the end of the year raised only one negative point. The Ampex VTR was going to be a very expensive item when it went on the market, and something more appropriate than the cratelike wooden structure in which the components were then housed was needed to make it look worth the price. When this and other minor flaws were corrected, the company leaders confided, they hoped to give a surprise demonstration of the world's first viable VTR at the Chicago convention of the National Association of Television and Radio Broadcasters in April 1956.

The VTR team's next crucial test was a private showing for a few more top executives in February. Charles Anderson designed the Mark IV console to house the VTR, and the six men who had created the recorder shared a common wish: may its performance equal its appearance. Their anxiety mounted when the expected "handful" of executives turned into a full-scale collection of the Ampex leadership. The team members saw no sign of enthusiasm from the spectators during the early part of the program. For the finale, they taped a two-minute speech, rewound the tape, and immediately played it back. The room erupted into spontaneous applause and yells of approval. The relief and happiness of the development team were exemplified by two of its members—who had fought continually over the years—who grabbed each other in back-slapping bear hugs, tears of joy streaming down their faces.

Visitors from CBS, ABC, CBC, and the BBC arrived later to spoil part of the surprise element of the company's plan to unveil the VTR at the NATRB convention in April.

In time, an arrangement was made to show the still-unassembled new Mark IV model at a CBS affiliates meeting in Chicago a day before the convention's opening. The Mark III recorder would simultaneously debut before a press conference in Redwood City, California. The development group was considerably enlarged, and many major refinements were added during the final six weeks in work that accelerated without regard to normal hours. Fred Pfost, who averaged more than one hundred hours per week during the final month, was still improving the performance of the heads up to the morning before the Mark IV was to be shipped to the Midwest.

Mechanical problems with the Mark III that threatened to delay the Redwood City demonstration were corrected, but the CBS engineers in Chicago complained about picture quality and noise in the Mark IV. In the twenty-four hours before the scheduled Saturday demonstration, the Ampex people made further adjustments and also received a welcome supply of new videotape from 3M that provided the best pictures and sound they had yet seen. The performances by the Ampex videotape recorders in California and Chicago created a sensation. By the time the four-day Chicago convention had ended, Ampex Corporation had taken orders worth about $4 million.

Not expecting that kind of demand, Ampex was faced with the immediate problem of providing the sixteen hand-built machines it had promised to deliver quickly to the TV networks, as well as setting up systems that could eventually mass-produce VTRs. The company succeeded on both counts, and the world's first videotape recorder was delivered to CBS's Hollywood Television City, where it recorded a live broadcast of "Douglas Edwards and the News" from New York for a later West Coast showing on November 30,

1956. NBC broadcast its first VTR program at the beginning of the new year, and ABC began its taped programming that spring with the arrival of daylight saving time.

MARKET EXPANSION AND COMPETITION

The first professional Ampex videotape recorders cost approximately $75,000 each, and the company had sold about 950 of them worldwide by 1960. When this revenue was coupled with its thriving business in audiotape recorders, it was clear that Ampex was an aggressive company on the move.

There was no slow-down in the innovation process as Ginsburg and his team continued to perfect videotape recording. Alex Maxey had started working quietly on his own in 1955 to simplify the process and built an experimental machine with a single rotating head by the end of the year. From that very crude recorder, Maxey completed a prototype of the first helical scan recorder in late 1956 and filed a patent application for it in March 1958. The company began high-volume VTR production with the Ampex VR-1000 in 1957; introduced the Ampex Amtec, which improved videotape interchangeability with the first time-base corrector in 1960; and produced the first electronic videotape editor, the Ampex Editec, in 1961. Ampex also introduced the first commercial helical scan VTR using 2-inch tape in 1961, the same year that similar recorders were made by RCA and Japan's JVC, Sony, and Toshiba.

Ampex had already felt the effects of competitive pressure from Japan in the field of audiotape recorders and had sent representatives to Japan in the late 1950s in an attempt to open markets and secure valid protection for its VTR patents. It had little luck in either quest, and the Americans soon had what they considered proof that the Japanese were pirating their concepts in the form of videotape recorders that seemed to be exact replicas of the Ampex VR-1000. In 1960 the company negotiated an agreement that put it into a joint venture with Sony to build a transistorized VTR, and won the right to license its patents to other Japanese firms. The Sony agreement degenerated into a legal squabble that lasted for nearly six years. Sony produced its first VTR within months of signing the one-page agreement and went on to become Ampex's number-one rival.

The seventy-year-old Alexander M. Poniatoff elected to retire in 1962. The company the Russian-born electrical engineer had founded in 1944 had become—with Polaroid, Texas Instruments, and others—what Wall Street called a growth company, a glamour stock. Poniatoff was replaced as CEO by William Roberts, who made it abundantly clear that he was intent on furthering that image. Under Roberts's guidance, Ampex began building new plants and enlarging existing ones in California, Colorado, Alabama, and Illinois while overseas operations were also being expanded. During the first nine years of Roberts's tenure, the work force grew from less than five thousand to more than fourteen thousand and company literature proclaimed that 60 to 80 percent of its revenues were derived from products no more than three years old. Much of that looked good to stockholders, as did the corporate goal of an annual growth rate in sales and profits of at least 15 percent. But Ampex had entered a new era in which success would threaten internal focus.

Ampex had been content with its dominance of the high-scale end of the market in both video- and audiotape recorders. Their expertise in miniaturization through the transistor enabled several Japanese firms, particularly Sony, to develop good videotape recorders during the early 1960s that sold for only a fraction of the cost of the Ampex professional product. Sony's Akio Morita has said that his company intended from the beginning to produce a VTR that could be brought into the consumer's home for the convenience of time-shifting—recording TV programs and viewing them at a time of the individual's own choosing.

But it was a late-1963 demonstration of an incredibly inexpensive recorder from England that jarred American corporations into a closer look at the potential of VTRs for the consumer market. Cinerama, Inc., sponsored the demonstration of the experimental fixed-head, longitudinal model made by Britain's Telcan. The recorder's picture quality impressed few people and little was heard of it afterward, but the machines' planned $175 price tag incited media speculation that truly affordable VTRs for the home were on the way.

Ampex was also inspired to begin thinking about producing the new low-cost VTRs. Many Ampex engineers doubted that a good-quality longitudinal machine could be built, but the project to make a recorder called the VR-303 began under the leadership of Gus Grant. By November 1964, Alex Maxey and John Streets created an air-bearing scanner that might make single-head helical video recording both viable and economical, and that breakthrough signaled the beginning of an extremely complicated segment of Ampex history.

The new consumer products division in the Chicago suburb of Elk Grove Village, Illinois, had already posted an excellent record in the production of audiotape recorders, but Rein Narma had been given informal assurance when he became chief engineer at the new plant that his crew would have an opportunity to develop VTRs for the consumer market when Ampex entered that field. When work began in California on a longitudinal VTR, Narma's engineers created an unauthorized model that some say worked as well as the finished VR-303. Narma and three of his engineers flew to California to examine the Maxey-Streets helical scanner, carried some of the key components back to Illinois, and started work on their own version of a consumer VTR. Within months, without any direct help from California, the Illinois innovators found a way to lower the cost of producing heads from $100 to $10 each and built an excellent prototype of a single-head helical scan recorder. It was clearly superior to the California-made VR-303.

The achievement of the Elk Grove Village engineers was appreciated by the Ampex management, although it stimulated an unhealthy rivalry between the California and Illinois engineers and further spotlighted the failure of the costly VR-303 project. The Elk Grove team was authorized to speed refinements on its new VR-7000 so it could be featured in a head-to-head public comparison with Sony's new low-priced VTR at the Chicago Music Show in June. But management could not explain why it had allowed two separate divisions to build radically different VTRs aimed at the same market within a few months of each other.

The Ampex and Sony consumer VTRs both stimulated the imagination of the working press at the Chicago show. *Life* magazine ran a lengthly feature article that extolled the virtues of both machines and predicted that they might well replace home movies as an American pastime. By November, the Elk Grove team had six prototypes ready for further demonstrations. Roberts announced that prices for the VR-7000 would range from $1,095 for a tabletop model to $2,495 for the deluxe model with TV receiver and video camera, but consumer testing in Hawaii produced an unpleasant response. Those surveyed reported that the suggested price was too high, they found it too complicated to operate, and none liked the idea that it recorded only in black and white when TV programming was moving to color.

Hoping the early surveys would prove to be incorrect, Roberts told the business press the company expected revenues of $10 million from home VTR sales in 1966 and was confident that figure would grow to $100 million annually by the end of the decade. Ampex was shipping about four hundred VR-6000 and VR-7000 home models to dealers each month by the end of 1966, and it appeared then that Roberts's optimism might be justified. But the number of orders dwindled to about two hundred per month by mid-1977, and marketing surveys revealed that the majority of sales had been to schools, hospitals, and other institutions that used the recorders as audiovisual aids. Ampex controlled more than 50 percent of the more expensive 1-inch-tape market, but the cheaper 0.5-inch recorders from Japan were better sellers.

Sony and Matsushita had begun a strong drive to capture the low-priced American market with two-head recorders using 0.5-inch tape in mid-1966, and those companies and a number of other Japanese firms were offering machines using both size tapes and better pictures within a year. Ampex introduced the VR-7500 color VTR in mid-1967 and continued to improve the performance of its higher-priced machines. But the competitors were proving that the American public was primarily interested in recorders that were reliable, easy to operate, and inexpensive. The Ampex financial picture had a rosy *surface* glow, but the high cost of expansion had cut steadily into its research and development funds during the 1960s: from a high close to 10 percent in 1962 to less than 6 percent in 1968. Elk Grove was now responsible for consumer video products, while Redwood City concentrated on VTRs for the broadcast industry. However, unrelenting pressure from the Japanese demanded a complete reexamination of the Ampex game plan.

William Roberts selected a young executive named Richard J. Elkus, Jr., to make a detailed study of the overall situation—with a special analysis of consumer operations at Elk Grove Village that might define Ampex's ability to compete in the home video market. After more than six months of investigation and reasoned input from a broad spectrum of experts, from technicians to marketers to university professors, Elkus delivered an impressively thorough report on November 12, 1968, that called for the company to revamp its thinking to outdo its competition. Backing his arguments with documentation—including detailed analysis of the most successful low-cost models produced by six Japanese companies, IVC, and Phillips of the Netherlands—Elkus urged management to develop a revolutionary new recorder that would meet the needs of what he called the instantaneous response market. He called his vision the *Instacorder* and described its function as an "audiovisual scratchpad."

Elkus followed his well-received November report with another in December that gave more details about his small videotape recorder that would permit Ampex to do more than maintain parity with its competitors. He warned that the company would have to anticipate competing VTRs' gaining innovative advances, such as "automatic threading, cassette or cartridge loading," and insisted that the Instacorder would meet those standards as a new development for the audiovisual field. He strongly urged Ampex to make a definite commitment to the consumer VTR market and described the Instacorder as the first of a complete line of new products for the home. The report stressed that the Instacorder should be compact and lightweight, easy to operate with automatic controls, rugged and adaptable to varied settings, priced no more than $1,600 for the audiovisual market—ultimately one-third of that for the home—and use cartridges or cassettes for easy loading. In less than sixty days, engineers at the Los Gatos, California, labs had built a working model of that recorder, and the company committed itself to all-out development of that system, which it had renamed *Instavideo*. Elkus was assigned the task of working out a complete business plan, and M. Carlos Kennedy was named Instavideo project manager at Elk Grove Village.

Elkus's business plan, released in May 1969, called for an estimated cash outlay of $1.5 million to meet the January 1971 target date for the introduction of the basic black-and-white Instavideo recorder. The company would introduce two new machines in the series—a monochrome model with stop/slow motion and editing features and a color version—in the second and third years of the program. Elkus expected Instavideo to produce $7 million in revenue in fiscal 1974. Optimistic as they were, his projections seemed entirely in line with all the market trends, but Ampex had never before attempted a crash program on that scale. Mass-producing the recorder would create special problems.

In very short order, the innovators were able to perfect Instavideo prototypes that were more advanced than any previous small VTR. Sony had just introduced its U-Matic (VCR) to the United States, but the less bulky and more versatile Instavideo was received with even more enthusiasm when it was introduced at a number of trade shows and conventions. The Ampex model weighed less than sixteen pounds and could be loaded automatically by using a tape cartridge that was 4.5 inches wide. The tape itself was compatible with other 0.5-inch recorders. The proposed price was competitive with the $1,000 Sony charged for its U-Matic. The Ampex monochrome recorder-player would sell for $900, a player only for $800, and a color recorder-player for $1,000. A 5-pound video camera could be added for another $500.

In what seemed to be a practical move, Ampex arranged for Japan's Toshiba to mass-produce the Instavideo in its new Toamco plant, but a series of problems—including language barriers between Ampex and Japanese engineers—kept the production lines silent as the January 1971 start-up passed. Despite Ampex's outward optimism and confidence, its inability to begin production would have long-term effects on the entire industry. Companies everywhere were debating whether to follow Ampex and produce cartridge machines using the EIAJ-1 standard tape format or switch to cassettes and alternative tapes.

The public explanation of the delays—working the bugs out of the production system at the Japanese plant—would continue to be used for more than a year, but, by the middle of 1971, it was no longer possible to hide the fact that Ampex was in deep financial trouble. The fiscal 1971 statement reported a $12 million deficit and William Roberts resigned. Work on the Instavideo program was temporarily halted in the expectation of fiscal 1972 losses of $40 million, but the status became permanent in October, when the program was abandoned officially. Henceforth, Ampex would supply only videotape for the home market and concentrate on the design and production of professional-quality videotape recorders for the broadcast industry.

Instavideo's death notice ended America's best opportunity to profit from a lucrative new market its innovators had helped create. Other U.S. firms would try, but all would fail.

BUILDING THE HOME VIDEO MARKET

Partially inspired by the excitement raised by Instavideo, a newly formed U.S. company stepped in to fill the void left by its demise and produced the only American-made video-cassette recorder to reach the marketplace. The new Cartrivision machine had both record and playback ability and created considerable interest when it debuted as a home product in 1972, but it was relatively primitive by Ampex and Sony standards and was rushed to market with inadequate planning.

Cartrivision's only model, a console with a built-in color TV receiver, cost $1,600, and that price tag put it out of the reach of most American families. Anyone capable of spending that much on home entertainment most likely had at least one color TV set already, making the costly receiver superfluous. With its recording capacity, Cartrivision was already ahead of several other systems being developed in the United States, but that feature was lost in the hyperbole about its wonderful ability to bring movies, Broadway plays, and other special entertainments into the home. The company tried gallantly to supply its customers with high-quality rental tapes but, in doing so, bowed to a film industry demand that negated another of the machine's most important features. Only the rental agencies had access to a device that would allow the tapes to be rewound once they were started. That meant the Cartrivision owner was forced to stay glued to his easy chair without as much as a bathroom break, giving him about as much control over his own appliance as he had over the projector at a movie theater. The Cartrivision company soon folded, with heavy financial losses.

Major American corporations such as RCA, CBS, and GE all tried to develop consumer-oriented video recorders but never succeeded in taking one to market. The one constant in the failure of American industry to produce a VCR for consumers was the belief that Americans could find little on television they would care to record and keep. Former CBS President Frank Stanton admitted as much when he said the lack of desirable software was a key factor in his company's decision to drop its costly EVR project. Some of the blame for the American loss to Japan in the race for home video success must be placed at Hollywood's doorstep. Historically fearful of new competition, the moviemakers actively fought the technology at first and only bowed to the inevitable after the Japanese had effectively cornered the VCR market. Video sales have since become the film industry's favorite weapon for salvaging the losses from box-office disasters.

RCA, the last American corporation to abandon its hope of winning a substantial share of the home video market, tried early on to remedy the entertainment quotient of the home entertainment product. Frank McDermott, once president of Hollywood's Four Star Productions, was hired by RCA and given multimillion-dollar budgets to tie up the rights to scores of motion pictures and other entertainments, before the company had made a firm commitment to a system on which to play them. Because of its long and successful history as a pioneer in professional and consumer products in radio, television, recording, and the like, many had expected RCA to develop the first VTR. The company built a working model video recorder while the Ampex program was still in its infancy, and technicians at its several state-of-the-art labs continued to make significant contributions to videotaping technology: the first color VTR, the first all-transistor professional VTR, and the first Quadruplex cassette spot recorder, among others.

But high-level indecision and an "embarrassment of riches"—both technical and financial—doomed RCA's home video program to failure. The complexity of that lengthy project can hardly be covered in a few paragraphs, but the consensus is that indecisive management, rather than a lack of innovative work on the engineering side, led to its sorry conclusion. RCA's size and wealth bred problems of interlab rivalry and wasted effort. One RCA team had a videocassette recorder that seemed the equal of any Japanese make ready for final development by late 1977, but corporate funding for the project was withdrawn in 1978, reportedly because Chairman Robert Sarnoff personally favored the potential of the videodisc. But when RCA's Selectavision Videodisc was finally unveiled on an NBC closed-circuit telecast in 1980, Sarnoff had been replaced and the introduction was handled by a new RCA chairman. And when the program was later abandoned, at a staggering financial loss, yet another RCA chairman made that announcement.

RCA's Selectavision Videodisc had a great deal going for it that could have made it competitive. The disc player was reasonably priced and produced an excellent picture, the videodiscs themselves were attractively packaged and featured a wide range of entertainment options, and RCA spent a fortune advertising the product. But the RCA machine could only *play* programs—it could not record. Hundreds of thousands of Japanese VCRs that could do *both* had already been sold in the United States, and motion picture studios were by then freely releasing movies for videotapes that could be bought or rented. There was no chance to overcome the VCR's head start. After twenty-six years of research, RCA had brought out the wrong home video product—much too late.

*   *   *

American businessmen, in general, were not prepared to match the long-term patience and foresight of their Japanese counterparts. The Ampex engineers who created much of the basic technology opened up the possibility of today's multibillion-dollar home video market, but the world has learned that it cannot dismiss Japan's electronics experts as mere "copycats." Consider these outstanding contributions by the Japanese innovators: the first viable single-head VTR by Toshiba in 1958, the first two-head recorder from JVC in 1959, the first Japanese professional transistorized VTR from Hitachi in 1962, the first video recorder for the consumer market from Sony in 1964, and the first small portable VTR from Hitachi in 1965. Engineers at Matsushita developed azimuth recording, which eliminated undesirable talking from adjacent tape tracks by a method of slanting the recording heads. Sony's engineers were the first to use azimuth recording to achieve a color video signal and also perfected the videocassette.

Sony deserves special recognition as the VCR pioneer. It sold more than 200,000 U-Matics in America before introduing its Betamax, the first VCR designed exclusively for the home, in 1975. Despite a slow start and predictions that it would fail, Betamax sales eventually began to soar. Ironically, an innovation developed by another Japanese company finally knocked Sony out of its dominant position in the American VCR market. About 100,000 Betamaxes had been sold when JVC introduced its rival Video Home System (VHS). The VHS cassette permitted eight hours of recording time compared to Beta's five and, although Betamax sells well, VHS now dominates the marketplace.

A study of the innovators who developed video recording provides a classic example of innovation's propensity for building companies and creating industries. It also offers a clear-cut and invaluable business lesson: inability to manage innovators and the products they produce—from drawing board to marketplace—can transform economic opportunity into financial disaster.

11

T-PA:
Biotechnology Versus
Heart Attacks

"The Genentech mission is to build a major independent, creative, and profitable business by applying innovative science to the development of new health care products."
—GENENTECH *Statement of Corporate Policy*

Earlier, in chapters 3 and 8, we examined miraculous new drugs and the processes that produced them at two of the world's largest pharmaceutical companies. A study of the drug Activase, Genentech's recombinant tissue plasminogen activator (T-PA), allows us to investigate a smaller firm's revolutionary new method of drug discovery and the innovators who found a way to bring it to the marketplace.

Very few discoveries in medical history have inspired the advance excitement of the genetically engineered T-PA, which has demonstrated an ability to dissolve the blood clots of killer heart attacks. The distinguished physicians who administered T-PA during the clinical testing became its staunchest advocates, and the communications media avidly reported their enthusiasms. Activase, its commercial name, had an eager, ready-made market even before it won the approval of the federal Food and Drug Administration in November 1987.

Recombinant T-PA was all the more remarkable because it was created by a company that had not even been in existence eleven years earlier. Genentech Incorporated, the biotechnology firm that achieved the breakthrough, has been a model of innovative techniques in research, marketing, and general outlook since its founding in 1976 by a twenty-eight-year-old visionary named Bob Swanson. Swanson earned a B.S. degree in chemistry at the Massachusetts Institute of Technology and stayed on to add an M.B.A. He joined a San Francisco venture-capital firm after graduation and was relatively content in that business until his lifelong interest in science caused him to pick up a book that detailed the story of the discovery of a structure of the DNA molecule by two British researchers at Cambridge in 1953.

Swanson says that book changed his life. With his own background in chemistry and business, it seemed to him that biotechnology could produce enormous benefits for mankind if properly utilized in a business setting. If a commercial biotechnology company could be organized—one that created and produced gene-spliced products capable of commanding the body's own mechanisms to prevent or correct natural faults—a new era in preventive medicine would be launched. He spent a considerable amount of time studying his idea and found that it appealed to him from every point of view. Commercial biotechnology was a scientific frontier that demanded exploration, and it was a business venture that offered enormous potential. He began a one-man crusade to make it happen.

Most of those he approached as prospective financial backers warned him that innovations always took time to pay off and biotechnology was so new that its large-scale commercial application was perhaps twenty years in the future. It became clear to Swanson that he needed the open and enthusiastic support of someone with scientific credentials far superior to his own, and he eventually found it in the person of Dr. Herbert Boyer, a renowned genetic engineer and professor of biochemistry at the University of California at San Francisco (UCSF). Boyer was not particularly impressed by the scheme when first contacted and warned Swanson that he could spare only ten minutes for their interview in January 1976. But the meeting went on for three hours, and, before it ended, Boyer found himself admitting that no insurmountable scientific barriers existed to make the plan impractical. Although reluctant to leave his university position, the professor agreed to serve as an unsalaried consultant and became a co-founder by investing $500 of his own money. That sum, equaled by Swanson's $500, was Genentech's entire working capital when the company was incorporated in April 1976.

Swanson's former employer, the venture-capital firm of Kleiner, Perkins, Caulfield & Byers, raised approximately $100,000 to finance the new company's first nine months of operation. Until substantially more money to enable Genentech to set up its own laboratories was raised, Swanson contracted the firm's work out to scientists at UCSF and the City of Hope. Even with that rather amorphous work arrangement, Genentech was able to claim a major success within a year with the cloning of somatostatin, the first human protein ever produced in a microorganism. This achievement attracted considerably more investment money in 1977 and allowed Swanson to hire his first two full-time scientists. The cloning of human insulin the following year brought greater financing, enabling the staff to be enlarged to twenty-three and the construction of Genentech's headquarters in south San Francisco to begin.

The discovery of the DNA structure by James Watson and Francis Crick enabled other scientists to develop the technology that inspired Swanson and Boyer to start a company that intended to produce pharmaceutical products virtually identical to substances found only in the human body. The term *gene-splicing* grew out of the discovery that the molecule deoxyribonucleic acid (DNA) contains a genetic code that controls much of the body's function. It acts rather like the type of information tape that is sometimes used to control the operations of automated factories. Perforations in the plant tape are coded information that tell the robotic machines the tasks they must do. DNA performs a similar function in the body, containing genes that supply coded information that dictates the form and function of each cell. Genes determine our physical makeup—color, size, and so on—and instruct the cells to produce the proteins that will enable them to function normally.

Recombinant DNA technology, or gene-splicing and genetic engineering as it is often called, was a process that was unimagined before Watson and Crick's 1953 discovery. In simple terms, that process involves joining genes from human cells with the DNA of another type of "host cell." Proteins are the primary product of all cells, and the ability to insert DNA segments carefully into the DNA of a host cell, with instructions to treat it as its own and follow the coded commands to produce a desired protein, offers medical science a means of replenishing supplies of the natural proteins. Inadequate levels of the vital proteins result in disease and disorders, and using biotechnology to reproduce vast amounts of the proteins for commercial purposes was the gist of the Swanson-Boyer goal when they founded Genentech.

Settled into their own laboratories in south San Francisco by 1979, Genentech's scientists were not yet aware of research being conducted in Europe that would ultimately have a momentous impact on the future of the company. Dr. Desire Collen and colleagues of the Center for Thrombosis and Vascular Research at the University of Leuven in Belgium spent 1979 investigating a natural protein called tissue plasminogen activator (T-PA) that was believed to be a natural dissolver of blood clots. The protein had first been discovered in 1947, but a lack of proper equipment prevented its being isolated or properly characterized. The T-PA that Dr. Collen was studying came from a unique cell line that had been supplied the previous year by a New York medical school. Called the *Bowes melanoma cell line*—obtained from an American patient named Bowes in 1974— it had been widely circulated among researchers because it secreted large amounts of T-PA activity.

Dr. Collen and his associates were able to purify the T-PA in autumn 1979 and scaled the purification process upward to produce about 2 grams of the protein. This made it possible to characterize its biochemical, biological, and physiologic properties systematically and enabled Collen to begin a lengthy series of in vitro tests. He was particularly fascinated by the activator's affinity for fibrin, the protein from which blood clots are composed. Even more revealing in vivo testing in rabbits proved that T-PA purified from the Bowes melanoma cell line was able to break up blood clots.

Still unaware of the T-PA research in Belgium, Genentech's growing staff of scientific innovators had compiled an impressive string of genetic engineering successes by 1980—somatostatin, human insulin, alpha interferon,

growth hormone—and the company's public relations people had made certain that the world was aware of these
accomplishments. Swanson decided to take the company
public in 1980, and Genentech stock was first offered on the
morning of October 14 at $35 per share. It soared to $88
within the first twenty minutes and closed at 71¼, setting a
stock exchange record for the greatest first-day advance of
an initial public offering. But an even greater economic
breakthrough would have to wait until Genentech established contact with Dr. Desire Collen—and T-PA.

Collen's research at Leuven was so successful by 1981 that
T-PA was tested on human patients for the first time. Its
use as a treatment for blocked renal veins in two Belgian
kidney transplant patients produced inconclusive results,
probably because of insufficient dosage. Dr. Collen made
the first formal report on his T-PA research that year at the
Fifth Congress of Fibrinolysis in Malmö, Sweden, and, in
one of those strange happenstances that so often occur in
scientific developments, that report led to Genentech's eventual involvement. Diane Dennica, a member of the Genentech Molecular Biology Department, was vacationing in
Malmö at the time and decided to attend the meeting to
increase her knowledge of fibrinolysis. Her colleagues say
she was not ejected from that private meeting, "because
she's young, pretty and petite, and they probably thought
she was the daughter of a member." After hearing his report, Ms. Dennica immediately contacted Dr. Collen.
     Within a few months, after long-distance communi-
cation between Leuven and south San Francisco and a Collen visit to Genentech, a collaborative agreement was made
to allow the American company to attempt to clone and
synthesize a T-PA gene from Dr. Collen's Bowes melanoma
cell line. The Genentech scientists were able to identify the

genetic sequence of T-PA before the end of 1981 and successfully expressed or synthesized enough of a recombinant
T-PA the following year to make it available for initial testing.

What the California researchers determined was that
tissue plasminogen activator is composed of a combination
of the basic amino acids, joined together in a long chain
called a polypeptide. T-PA consists of 527 of these amino
acids arranged in a specific sequence. To produce recombinant T-PA, the genetic engineers copied the instructions
for the T-PA polypeptide chain from a human cell and in-
serted the gene into a plasmid, or small circle of DNA. They
then cultured the single cell in a medium containing liquid
and nutrients for laboratory-scale fermentation under the
best possible conditions. Results of Genentech's successful
program were reported at the Sixth Congress on Fibrinolysis
at Lausanne, Switzerland, in 1982, and Dr. Collen was given
enough of the recombinant protein (rT-PA) to begin animal
testing.

The early testing by Dr. Collen's team in Belgium
demonstrated that the activity of Genentech's recombinant
T-PA in rabbits was indistinguishable from that of natural
T-PA. The University of Leuven group had already supplied
natural T-PA for wider animal testing in a program that had
originated in a workshop of the American National Institutes
of Health at the end of 1981. That cooperative effort between the Belgium group and Dr. Burton E. Sobel's team
at St. Louis' Washington University had demonstrated the
natural protein's ability to break up blood clots present in
acute myocardial infarctions (heart attacks). When coronary
thrombosis was induced in dogs for the trials, the natural
T-PA effectively dissolved the blood clots quickly, without
causing other copious bleeding, and showed a remarkable
ability to act on only the fibrin of the damaging clots. After

Genentech had created small-scale supplies of rT-PA and won approval for its use in the same program, its gene-spliced protein demonstrated identical beneficial results in both dogs and baboons.

Positive news of the animal trials and the possibility of imminent testing of its recombinant product on human patients was joyfully received in the offices and laboratories of Genentech Inc. If natural T-PA and its clone proved to be equally quick and effective in dissolving blood clots in human heart attack victims, it would be a miraculous breakthrough capable of saving many thousands of lives each year. Trials of natural T-PA on human patients were scheduled to begin in February 1983; if they produced positive results, testing of the recombinant T-PA would follow. Genentech geared up for increased-scale fermentation of rT-PA and prepared new and improved methods for product recovery and purification that would ensure adequate supplies for those tests.

The natural T-PA clinical trials with patients suffering from acute myocardial infarctions, conducted by Dr. F. van de Werf and his group at the University of Leuven and by Dr. Sobel and his team at Washington University, began in February and continued through September. Administered intravenously, the natural T-PA emphatically demonstrated its efficiency in dissolving deadly blood clots from the arteries of six of the first seven heart attack victims to whom it was administered—within thirty to sixty minutes and without evidence of any harmful side effects.

That impressive demonstration of the efficacy of Dr. Collen's natural T-PA made the doctors all the more eager to learn whether recombinant T-PA could produce equally promising results. The FDA approved blind, randomized trials that would be conducted at Washington University, Massachusetts General Hospital, and Johns Hopkins University with Drs. Sobel, Herbert Gold, and Myron Weisfeldt heading their respective teams. The stakes were high, and the possible rewards, both humanitarian and economic, were mind-boggling.

TESTING, CONTROVERSY, AND APPROVAL

The earlier trials with the natural T-PA demonstrated that the protein was most effective in breaking up blood clots when it was administered as quickly as possible after the first symptoms of heart attack appeared, and Genentech's recombinant T-PA was tested in the same manner on fifty volunteer patients at the three medical centers. The gene-spliced protein was administered intravenously in doses of 0.5 milligram per kilogram of body weight over a sixty-minute period or, in some instances, that dosage was infused for the first hour with 0.25 milligram per kilogram of body weight added during the second hour. The rT-PA opened the occluded coronary arteries of 75 percent of those treated—and evidence of undue bleeding or other negative side effects was either absent or minimal in most patients.

The strongly positive results of the first clinical tests of rT-PA produced unanimous appeals from the trial doctors in St. Louis, Boston, and Baltimore for the early start-up of more thorough clinical trials. Their formal requests received a speedy response, and the earlier testing served as a foundation for the new trials that would be conducted in both the United States and Europe. The American testing—Thrombolysis in Acute Myocardial Infarctions (TIMI)—would begin almost immediately under the sponsorship of the National Heart Lung and Blood Institute (NIH). The European Cooperative Trial—to be conducted

in Belgium, France, the Netherlands, and West Germany—
would begin in July 1984.

The NIH conducted an initial open-label phase of
testing in eighty-seven American heart attack sufferers in
the first two months and then proceeded on to a double-
blind direct comparison of the effectiveness of the rT-PA
with another thrombolytic agent, streptokinase. The European trials had started off with direct comparison testing
with streptokinase. The first phase of the European trials
continued until December 1984, but the larger U.S. program
continued on into the new year. No reports would be published for some time, but Genentech's people, working
closely with the NIH, were aware that their product was
faring very, very well in the comparison testing with streptokinase. This was a considerable source of relief for the
young biotechnicians because streptokinase had been approved by the FDA in 1982, less than a year after Genentech
had started its T-PA research.

To the surprise of most of the medical establishment,
the Policy Advisory Board of the TIMI study group issued
a recommendation to the National Institutes of Health that
phase I of the clinical testing be ended in February 1985,
somewhat earlier than planned. And this decision brought
an unexpected and most unusual boost for Genentech's recombinant T-PA. The distinguished *New England Journal
of Medicine,* stressing that it was not in the habit of publishing preliminary or inconclusive reports on a new drug
before clinical trials had been concluded, made an exception
in the case of rT-PA in its issue of April 4, 1985.

In explaining its unusual publishing action, the *Journal* fully recounted the drawbacks of the already approved
streptokinase and stated its belief that these were not present
in the rT-PA. Both agents were most effective when administered as quickly as possible after the first symptoms of

heart attack, but streptokinase's maximum effectiveness relied on its application through intracoronary infusion—direct insertion into the heart via a catheter—in a procedure that was time-consuming, complicated, and expensive. The precious time element had caused many doctors to use streptokinase intravenously before the trials, and this had prompted the direct comparisons between streptokinase and rT-PA in intravenous applications. And these comparisons in the first phase of the TIMI trials, as reported by the prestigious medical journal, clearly showed rT-PA to be far more effective than streptokinase.

"When administered in what the investigators believed to be optimal dose," the *Journal* reported, "intravenous tissue-type plasminogen activator was clearly superior to streptokinase in its immediate thrombolytic effect. Sixty-six percent of the patients given the plasminogen activator had a prompt and significant improvement in perfusion of the obstructed coronary vessel, whereas similar effects were observed in only 36 percent of those treated with streptokinase."

That sensational bit of news would have a profound impact on the medical community and much of the general public. The report that the new gene-spliced product was nearly *twice as good* as the only available remedy immediately sent Genentech's stock soaring, literally as well as figuratively. *Lancet*, the British medical publication, added more luster to the recombinant agent's reputation when it published the results of the European Cooperative trials. The report confirmed that rT-PA had demonstrated that it was far more effective than streptokinase in the 129 patients treated in Europe.

Phase I of the TIMI trials had been far too limited to produce conclusive findings about the efficacy of rT-PA, but their results were sufficiently important to prompt the

National Institutes of Health to speed the start of Phase II. The second phase would require more time and involve many more patients and medical centers to determine whether the genetically engineered medicant was both safe and beneficial as those terms were defined by the FDA. Confident that their product would pass its first test, Genentech's innovators had already created methods that would allow them to produce greater supplies of rT-PA.

In producing the original small-scale supplies of rT-PA, the scientists had used what they termed the roller bottle method in the laboratory. When it became apparent that commercial quantities would be needed, the innovators went to another manufacturing process called suspension culture. The upscale method caused an enzymatic reaction in the substance that produced a single-chain configuration in place of the double chain of the original rT-PA. The natural T-PA found in the body is single-chain so, by serendipity, the scientist had made the newer product closer to what is normally found in the body. It was otherwise identical in every way, and the company expected no significant changes in the results it produced.

Phase II of the NIH clinical trials continued to go very smoothly and Genentech felt enough confidence in its recombinant product to file its first product license application with the FDA early in 1986. Testing with the original double-chain rT-PA was producing excellent results with a dosage of 80 milligrams, but as the trials continued it became clear that the new single-chain product was somewhat less effective unless the dosage was raised. A level of 100 milligrams would prove to be the optimum level, but, because determining proper dosage is a primary reason for clinical trials, some of the participating doctors attempted to push the level higher in the hope of attaining even better results. Those who raised the dosage to 150 milligrams found that rT-PA quickly dissolved the coronary blood clots, but it also seemed to induce brain bleeding in 1.2 to 1.9 percent of the patients treated. That was not altogether surprising because T-PA's natural function is the breaking up of blood clots that would permit a better flow of blood, but incomplete reports of the incidents did raise alarms in some quarters. The 150-milligram dose was clearly too high and, when the dosage was brought back down to the 100-milligram level, virtually all evidence of undue bleeding ceased.

It took more than a year for the FDA to take any direct action on Genentech's New Drug Application, but the company's sustained effort to inform the medical community of rT-PA's potential brought it strong support from many sectors. More and more articles about the new miracle drug that could reduce deaths from heart attacks began to appear in the public print. By spring 1987, the company's stock, after several splits, had jumped back up to $65 per share in anticipation of imminent FDA approval. It seemed a good bet for speculators because the agency was reportedly ready to approve streptokinase for intravenous application, and the early testing had convinced NIH researchers that the Genentech product was twice as effective. But controversy exploded in May when the FDA approved streptokinase but refused to bestow the same blessing on rT-PA.

The cardiorenal advisory committee, which had the power to recommend approval or rejection, forwarded neither option to the FDA after its formal May 29 meeting on the recombinant clot dissolver. It advised the FDA to withhold approval until more information was supplied, citing concern that an insufficient number of patients had been tested at the 100-milligram dosage to allay the fears about excess bleeding. The advisory committee also wanted more proof that rT-PA provided patient benefits over and above clot lysis. The recommendation to postpone approval inspired an unprecedented hue and cry in the medical community and the media. Editorials in major newspapers— *The Wall Street Journal* ran five—suggested the FDA was being callously overcautious in delaying the availability of a wonder drug that had demonstrated its ability to save the lives of heart attack victims. The genetically engineered agent continued to receive virtually unanimous praise from doctors everywhere.

Although disappointed, Bob Swanson and his Genentech staff considered the advisory panel's actions only a temporary setback and set out to work even more closely with the FDA over the summer of 1987. In truth, the additional information the committee had wanted in May was partially collected at the time but had not yet been released by the NIH participants. The completed testing brought proof that the recommended dosage of 100 milligrams resulted in intracranial bleeding in less than 0.5 percent of those who received the therapy. And when scientific data were compiled from separate clinical studies in Maryland and Australia, rT-PA had demonstrated its ability to improve ventricular function and limit heart muscle damage in those who suffer myocardial infarction.

The FDA gave rT-PA its formal approval just six months after denying the original request, causing some to argue that the FDA had "caved in" under the unusual public pressure. The approval came on November 13, 1987, and Genentech announced that it would market rT-PA under the trade name Activase. The FDA approval included the customary list of precautionary alerts. Doctors were warned that Activase was not recommended for patients who had a high risk of hemorrhage, a history of stroke, bleeding disorders, active internal bleeding, recent surgery, or severe and uncontrolled high blood pressure.

Genentech's close cooperation with the federal agency expedited the licensing procedure, and the aggressive young company was able to rush its product to market in record time: exactly one week after the first appearance of streptokinase that had been approved six months earlier.

INNOVATIONS IN MARKETING

When Genentech's leadership returned to California from the FDA hearings in Washington, the stage was already set for a huge victory celebration. Fireworks illuminated the sky over the company parking lot and a giant tent where the entire work force had gathered for an Activase celebration demonstrated the freewheeling spirit of Genentech's corporate culture. Old-line pharmaceutical firms might have toasted the approval less publicly or waited for proof that they had a commercially successful product, but the young biotechnology workers had no such reservations. They were completely confident their gene-spliced "clot buster" would be a commercial sensation because their marketing and public relations staffs had been successfully promoting the benefits of tissue-type plasminogen activator for years.

Genentech's energetic and innovative program to win scientific approval and financial support for its experimental drug was a revolutionary approach that disturbed some members of the scientific community. One pioneering expert on clot-dissolving agents suggested publicly that Activase would never measure up to the promises of its publicity campaign and added that the only "miracle" about it was the DNA technology that produced it. But his was distinctly a minority opinion, as rT-PA was accepted by an overwhelming majority of the medical establishment as a true miracle breakthrough drug. It is widely believed that physicians were among the heaviest early investors in Genentech stock.

The young company used many innovative ideas from the beginning to assure a positive reception for the product it had created through DNA technology. Working as closely as possible with the National Heart Lung and Blood Institute, Genentech was able to encourage participation by the outstanding individuals and organizations in the field. Before the clinical trials ended, every major teaching hospital in the United States was involved, along with nearly every major cardiovascular group. The company began sponsoring symposiums on methods of dissolving cardiac blood clots during the first year of NIH trials, rallying even wider support from the medical community. Particularly helpful were the events timed to precede the start of the annual meetings of the American Heart Association. The discussions on thrombolysis in acute myocardial infarctions were research-oriented and not limited to the efficacy of Genentech's own agent, but reports on rT-PA's early success prompted the nation's leading cardiologists to make it the center of attention without undue encouragement from the sponsor. The specialists who participated in the NIH trials—and many of those who became privy to early information about rT-PA because of the Genentech symposia—became its staunchest supporters.

FDA approval sent Genentech's marketing and PR teams into phase II of their strategy to make Activase the agent doctors reached for automatically when treating the victims of heart attacks. Less than three weeks after the FDA had given its formal approval, Genentech organized interactive teleconferences in hotels in twenty-two American cities. The videotapes were quickly distributed to more than five hundred hospitals and the company claimed that thousands of doctors across the country had seen the videos and/

or received personal briefings on Activase within a few months of its release.

Some have questioned the early assertion that T-PA is *twice* as effective as streptokinase, but a year after entering the marketplace Activase was the overwhelming favorite of American cardiologists, despite its higher cost. A patient paid $200 for a treatment with streptokinase in the autumn of 1988, when a 100-milligram course of T-PA therapy set him back $2,000. But Genentech's concerted effort to spread the T-PA news had made a strong impact on potential patients as well as doctors, and the price factor diminishes when human life is threatened. Genentech strongly defended its Activase pricing throughout 1988 but faced increasing pressure from outside quarters.

Private insurers agreed to pay for T-PA therapy when the FDA approved the clot dissolver, but hospitals that treat large numbers of Medicaid patients were on less firm ground. The panel that counsels the U.S. Congress on Medicare matters—adjustments for new technology, changes in the health-care market, inflation, and so on—conducted its own study comparing Activase and streptokinase on largely economic terms and, in January 1988, advised the government to reimburse hospitals for only one-half the cost of T-PA therapy. No final decision had been reached when this book was being prepared, but the American Society of Hospital Pharmacists predicted that U.S. hospitals could lose between $225,000 and $1 million in 1988 on unreimbursed T-PA costs.

In the meantime, Genentech Inc. has demonstrated that it intends to be a vigorous defender of its patents and the market it has constructed. The company was awarded a U.S. patent in November 1987 for its fundamental genetic engineering process, and that all-important patent was followed by one issued in Britain in June 1988 that gave Genentech's ally, the University of Leuven, broad patent protection for T-PA. Legal experts say the field of biotechnology is so new and complicated that continual lawsuits will be necessary to establish key guidelines. Bob Swanson has said his firm has no interest in blocking others from developing gene-spliced products and is willing to license its patents at reasonable rates on a case-by-case basis. But Genentech's attorneys have been busy testing the strength of those patents in the courts. Immediately after the T-PA patent had been awarded in Britain, Genentech filed a suit against Burroughs Wellcome Inc., the American subsidiary of the British Wellcome Foundation, and the Massachusetts-based Genetics Institute Inc., which were reportedly ready to produce a T-PA agent of their own. The California firm also filed a suit against an affiliate of the Monsanto Corporation, claiming it was using Genentech's patented technology to produce another Activase-like drug.

ESTABLISHING THE CLIMATE AT GENENTECH

Some of the more orthodox students of the pharmaceutical industry have said that a company like Genentech "could only happen in California." This rather flip appraisal was largely built on the firm's well-known informality, its love of partying, and its youthful brashness. But its approach to the creation, development, and marketing of recombinant tissue plasminogen activator had made it evident that Genentech's offbeat corporate culture was thoughtfully structured to create a climate in which innovation could thrive.

CEO Bob Swanson makes certain that every Genentech employee receives the handsome blue-and-gold brochure, *Statement of Corporate Policy*, to ensure that no one ever loses sight of "what got us here." Insiders insist that

Swanson's personal enthusiasm and belief in the potential of biotechnology have filtered into every department of the company he co-founded with Dr. Herbert Boyer. Genentech has lost few members of its scientific staff, compared to other high-tech companies, and employees say this is because the management has made a conscious effort to establish a rare laissez-faire aura that recognizes the researchers' individuality while it encourages teamwork in the pursuit of common goals.

The almost-collegiate feeling of "togetherness," they say, is a key ingredient that allows them to feel they are on the leading edge of scientific research and moving steadily closer to the corporate goal of becoming the first biotechnology company to grow into a major pharmaceutical manufacturer. To further that aim, Swanson provided all the scientific staff with stock options, and, with the stock splits, many have since become millionaires. Close to 98 percent of the work force are now Genentech stockholders, with a vested interest in the company's success. The huge Activase victory celebration was no rarity. Regular parties—the company calls them "Ho-Hos"—are an established Friday night routine at Genentech. Swanson always attends, and scientists, salesmen, marketing and PR people, and employees on all levels drink beer and unwind in an atmosphere that fosters the we're-all-in-this-together attitude that the company cherishes.

Genentech has made remarkable scientific and financial progress since its unlikely start-up in 1976, and management claims it rewards its genetic engineers by offering "a rare blend of academic and entrepreneurial creativity where scientists are afforded a degree of freedom unusual in the pharmaceutical industry. Genentech's liberal publishing policy resulted in more than 750 papers in scientific journals and more than 3,500 patent applications being filed

THE INNOVATORS

in the first ten years of business." To encourage innovation further and keep its scientists on top of the latest developments, the company hosts many symposia for its researchers throughout the year and works closely with academic institutions around the world on postdoctoral programs and other collaborations.

In 1987, while the company was waiting for FDA approval of rT-PA, Genentech's researchers discovered twenty-eight new compounds or unique applications for known compounds through biotechnology. The company had two new anticancer agents that were in the testing phase as this book was being written, tumor necrosis factor (TNF) and gamma interferon, the latter also considered a promising treatment against viral diseases. Genetic engineering is well suited for developing vaccines, and Genentech is attempting to create agents to prevent hepatitis B and herpes simplex I and II. Other products being developed are Factor III, for treating hemophilia, and Relaxin, a natural hormone designed to ease childbirth.

Innovative techniques in the business offices as well as the laboratories have accelerated Genentech's success. Genentech, for example, became the first company ever to use a public research and development partnership to provide the financial resources needed to begin clinical testing on an experimental drug. The company had a market capitalization by 1988 that was 180 times its market earnings and 48 times the size of its direct product sales. More than $80 million, out of total 1986 revenues of $134 million, was reinvested in research and development. Revenues rose to $239 million in 1987, with $96.5 million going for R&D. And, thus far, growth seems to have had no negative effect on the company's culture or dedication to innovation. Today Genentech Inc. employs more than fourteen hundred people, who work in the 850,000 square feet of laboratory, office, and manufacturing space the company occupies in south San Francisco. Its two commercial products, Protopin and Activase, are manufactured in a 100,000-foot plant and the two earlier products, Humulin and Roferon-A, were licensed for marketing by Eli Lilly & Co. and Hoffmann-La Roche, respectively.

After a year on the market, Activase had firmly established itself as an effective aid for prolonging the life and health of those who suffer heart attacks. In its long journey from the body of a New York cancer patient to a Belgian research center and then back again to the United States for its remarkable rebirth in the laboratory of a California biotechnology company, tissue plasminogen activator has clearly become the first major product to be derived from genetic engineering.

12

Tomorrowland: Innovations in Our Future

"We work day after day, not to finish things, but to make the future better—because we will spend the rest of our lives there."

—CHARLES F. KETTERING

A multitude of innovations since the end of World War II have made our planet an increasingly global society and, by and large, a better and more efficient place in which to live and work. Ongoing advances now promise to push even some of our newer innovations toward obsolescence. Some of these new technologies are *on* the horizon rather than beyond it, meaning we have an opportunity to start reaping some of the abundance of new creative thought and effort during the final decade of the twentieth century—rather than in some fuzzy, indistinct future world.

Many factors will determine the eventual timing with

which these revolutionary innovations may reach the marketplace, and both money and resolve, as always, will dictate whence they come. But the prospects are as dazzling as the problems are perplexing. As we have learned from some of our earlier studies, inventors often have little idea of the varied uses that will be found for their scientific innovations or the dynamic impact they will have on modern technology. They are almost always appalled at the time required to adapt them to the basic commercial systems they were designed to complement.

The struggle for the horizon innovations we will examine in this chapter is producing similar human dramas now in laboratories around the world. The cast of characters and the subject matter differ, but the frustrations and determination are nearly identical. Not all of the exciting projects presented here will produce applications that will make their way into the home or office before the year 2000, but most of them have been the focus of intense study for years in the academic, governmental, and commercial research centers of the world, making the hackneyed prediction "It's only a matter of time" singularly appropriate.

SUPERCONDUCTIVITY

Few scientific papers have created the excitement or sent technicians scurrying to their workbenches as quickly as one published in 1986 by a Swiss physicist employed at IBM's Zürich research center. That paper by Dr. K. Alex Mueller described how he and his German-born colleague, Dr. Georg Bednorz, had put together a material that conducted electricity with absolutely no resistance at a temperature of 30 degrees on the Kelvin scale.

The announcement had little meaning for the average

person, but physicists recognized it as a monumental break-through in a search that had stymied the scientific community for decades, one that might permit rapid advances in a new technology that was capable of producing incalculable ben-efits for all of society. Indicative of the significance the scientific community ascribed to the Mueller-Bednorz dis-covery was the Nobel Prize for Physics the two shared in 1987, less than a year and a half after achieving their lab-oratory success.

The superconductivity story actually began in 1911 when Dutch physicist Heike Kamerlingh-Onnes, wanting to test the effect of extreme cold on various materials, lowered the temperature of expensive helium gas until it liquefied at a temperature that science calls absolute zero: about 459 degrees below zero on the Fahrenheit scale (-459 degrees Fahrenheit). Kamerlingh-Onnes placed a mercury wire into the liquid helium until it cooled to a temperature of -452 degrees, sent electricity through it, and saw that it flowed without any sign of the resistance normally present. Many thousands of experiments in the seventy-five-years between 1911 and 1986 uncovered hundreds of materials that were capable of superconductivity—at temperatures very close to absolute zero. As exciting as the prospects were, supercon-ductivity was impractical because of the prohibitive cost of liquid helium cooling. The challenge was discovering ma-terials that might superconduct at much higher tempera-tures.

Superconductivity occurs when some materials, cooled to a critical level, lose all resistance to the flow of electrons—microscopically minute particles that are nega-tively charged—creating what is more commonly called elec-tricity. In the standard materials used to conduct electricity, the electron flow is disrupted and part of its energy dissipated when some of the electrons inadvertently break loose from the mainstream and collide with the conductor's atoms, causing heat and resistance and dictating the need for some power source to keep the electrons flowing in unison.

Scientists were fascinated by the ability of many common metals to conduct electricity without resistance when brought down to extremely cold temperatures but had no clear idea of what caused the phenomenon or how it worked for nearly fifty years after Kamerlingh-Onnes's discovery. Interestingly, it was continuing research by one of the inventors of the transistor that provided the most acceptable answer. John Bardeen, the transistor physicist, and colleagues Leon Cooper and John Schrieffer won a Nobel Prize in 1972 (Bardeen's second) for their explanation in the BCS theory that bears their initials.

The BCS theory holds that lattice vibrations within the material through which the electrons pass affect their flow. At normal temperatures the lattice vibrations—the continual random movements of atoms that make up the conducting material's crystalline structure—are enough to shake electrons out of their flow and create heat, or resistance. As the conducting material is cooled, the lattice vibrations are correspondingly calmed until they reach a point where their movements become uniform, permitting the electrons to flow through unhindered.

It was this theory that enabled scientists to experiment with materials, primarily metals, in which the lattice vibrations might be controlled without resorting to extremely cold temperatures. But progress was very slow. Scientists the world over, doubtful that a room-temperature superconductor would ever be found, had set a temperature goal they hoped might be achievable: 77 Kelvin (the scale used to measure extreme cold) or −321 degrees Fahrenheit. Liquid nitrogen, which presented no handling problems and cost less than a supermarket cola, could cool anything down to 77 Kelvin. A material that could provide zero resistance at that temperature might make wide commercial application economically feasible. Bell Labs and Westinghouse each developed a superconductive material, niobium-3 germanium, in independent research in 1973 that set a record high superconductivity temperature of 23 Kelvin (−419 degrees Fahrenheit).

That high-temperature performance remained unchallenged for the next thirteen years, and, during that period, many research organizations abandoned the quest for superconductivity, their own practical and theoretical research having convinced them that 23 Kelvin was very close to the maximum temperature at which zero resistance could be achieved. If this was true, the prohibitive costs of lowering temperature to the point where superconductivity was possible would keep it simply an interesting laboratory phenomenon with little or no practical application.

The value of the niobium alloy's superconductivity was such that complicated insulating methods were devised to permit liquid helium cooling that made it viable in enormously powerful magnets that became integral parts of giant particle accelerators and even experimental Japanese trains. The most notable application of the superconducting magnets has been in the unique magnetic resonance imaging machine, which the medical establishment considers far more effective than the X ray in providing detailed images of the interior of the human body. But those few ultracostly projects were all that materialized and, as the years passed, the 1973 record of 23 Kelvin remained untouched and the 77 Kelvin goal seemed increasingly unrealistic. Pessimism faded, almost overnight, in 1986.

The Mueller-Bednorz breakthrough in superconductivity fits the classic mold of accidental discovery that makes scientific history even more compelling, although their so-called accident also grew from important related research. According to the Swedish Academy, the two were actually looking for something that might completely block the conduction of electricity. They had been deliberately experimenting with various ceramic compounds known to be poor conductors that were sometimes used as insulators when they came across one that caused them to change their course. Mueller and Bednorz were astounded to find that it showed signs of superconductivity. Physicists the world over had concentrated on metallic elements and metal alloys as the most likely superconductors. However, one of their ceramic compounds, a mixture of barium, lanthanum, copper, and oxygen, demonstrated superconductivity at 30 Kelvin: 22 degrees Fahrenheit warmer than the record set in 1973.

After repeating the experiment many times to make certain their readings were not in error, Mueller and Bednorz were convinced they had made an important discovery. Mueller published a paper that detailed the exact procedure he and Bednorz had followed in a German scientific journal in spring 1986, which enabled physicists elsewhere to conduct the same experiment. Scientists from the University of Tokyo were the first to duplicate the Zürich success, closely followed by American researchers, and the validity of the Mueller-Bednorz discovery was soon verified in labs all over the world.

The zero resistivity of the Zürich copper oxide material demanded a new approach to the search for superconductivity and served as a starter's gun for a race that may be unprecedented in scientific history. Every major research center around the world, in addition to a multitude of small labs everywhere, began experimenting with variations of the Mueller-Bednorz ceramic compound. The superconductive material made by the Swiss researchers was an *oxide*, a chemical compound containing oxygen, and this proved to be the primary reason for its success. Ironically, the fact that some oxides had a tendency to conduct better when chilled was not completely unknown even before Mueller and Bednorz made their breakthrough, but few had seriously investigated their potential for superconductivity.

The Zürich IBM compound of oxygen, copper, barium and so-called rare earth minerals (that are common rather than rare) was easy to make, and its components were all in plentiful supply. With the proper formula, a chemist could simply mix the ingredients, bake the powdered mixture in an oven overnight at 1,000 degrees Fahrenheit, regrind the baked mixture, and bake it again several times to make it more pure. The baked and purified powder could then be pressed and fired into a brittle ceramic material with a texture very similar to that of dinnerware.

Frenzied new interest in superconductivity, inspired by the achievement of Mueller and Bednorz, caused their high-temperature mark to be broken within months. By Christmas 1986, a number of research teams were reporting superconductivity in copper oxide materials at a temperature of −390 degrees Fahrenheit, a full 7 degrees warmer than the original Zürich success. Scientific journals around the world were inundated with reports of advances, and it was clear that the push for practical superconductivity was proceeding at full throttle. But few were prepared for the news of February 1987 that proclaimed a gigantic step forward.

Physicist Dr. Paul C. W. Chu headed a team of scientists at the University of Houston that was working on variations of the original formula created by Mueller and Bednorz. Chu's team was a small one that took pride in its innovative approach to problems. Wondering what would happen if the IBM compound were subjected to 10,000 to 12,000 times normal atmospheric pressure, the Houston experimenters were delighted to see that superconductivity was present up to 52 Kelvin when it abruptly fell off. Theorizing that the pressure enhanced superconductivity by compressing the compound's structure—but only to a point— Chu and his colleagues next tried to compress the structure internally by replacing various ingredients of the Zürich compound with elements that had smaller atoms.

Almost exactly a year after Mueller and Bednorz had made their breakthrough, Chu had a mixture of copper oxide, barium, and yttrium baked, pressed, and fired into a ceramic material that demonstrated zero resistivity at an astonishing high temperature of 93 degrees Kelvin: −290 degrees Fahrenheit. That was not only 100 degrees better than the previous best mark; it was superconductivity at 31 degrees Fahrenheit higher than the magic 77 Kelvin that many scientists had thought unattainable only a year earlier. By creating a superconductive material that could be adequately cooled by inexpensive liquid nitrogen, Chu had seemingly cleared the way for innumerable commercial applications of superconductivity.

Chu applied for a patent in January 1987 and refused to give details about his material until he published a paper in the *Physical Review Letters* on March 2. But even the word-of-mouth reports of the new breakthrough stimulated similar experiments in laboratories all over the world. By smashing through the long-sought 77 Kelvin barrier, the academic innovators at the University of Houston had removed superconductivity from the realm of pure research and made it a subject of extreme importance to all profit-making organizations. Others were en route to similar achievements even before the annual meeting of the American Physical Society that was scheduled for March 20, 1987, at New York's Hilton Hotel.

That meeting was expected to attract much attention because of its principal subject matter, superconductivity, and the panel of five experts who had been invited to speak, but few expected it to be turned into what some have since labeled "The Woodstock of Physics." Instead of five speakers, fifty-one individuals from a host of research centers asked permission to make presentations. The normal attendance of several hundred became an eager crowd of more than three thousand, and the annual meeting grew into a marathon session that lasted for more than seven hours, not breaking up until nearly four in the morning. It was evident the time had come when everyone wanted to stake a claim on some portion of superconductivity.

Illustrative of the rapid progress being made were the reports offered by scientists of Bell Laboratories. Hard on the heels of Dr. Chu's February paper, Bell's people announced that they had taken that work another step forward and produced a *single-phase* compound of yttrium barium copper oxide, meaning it could be represented by a single chemical formula, allowing the scientists to determine the specific crystal structure and detailed composition of the material and perhaps guiding them to other high-temperature superconductors. Bell's single-phase material demonstrated superconductivity at 91 Kelvin. Using a compound of barium, yttrium, europium, and copper oxide, the Bell researchers also announced success in experiments that would lead to powerful superconducting magnets that might be used in a wide variety of applications. They claimed the highest "critical field" levels yet announced at 77 Kelvin, which were comparable to the best magnetic field values obtained from magnets cooled by liquid helium. And they had also demonstrated the practicality of processing two different forms of superconducting ceramics, one a toroid or doughnut-shaped ceramic and the other a ceramic tape that was superconductive at a temperature "above 90 Kelvin." Reports of amazing advances continued throughout

1987, but, by the end of 1988, no viable room-temperature superconductor had been conclusively demonstrated. Most experts simply agreed that "It could come today, sometime in the next month, or maybe not even for another ten years."

Progress on new superconductivity materials has slowed because physicists have not yet discovered exactly how the new ceramics works. Not everyone agrees that the BCS theory of lattice vibrations is applicable to ceramics. Basic problems, such as the cessation of superconducting when too much current is sent through the new materials or when they are confronted by an overly strong magnetic field, have not been solved. Better knowledge of their structure and behavior will help determine how far their limits can be pushed.

Inordinate excitement about room-temperature superconductivity has disguised the progress made on superconductive materials that seemed viable for large-scale industrial use after the 77 Kelvin barrier had been breached. An obvious challenge has been the search for methods of molding the brittle ceramics into new forms that might allow them to perform their intended tasks. Prototypes of superconducting wire that can be formed into coils and other shapes have been made and some have demonstrated an ability to carry over one hundred times more current than was previously believed possible. IBM scientists discovered a method of "spray painting" superconducting material onto large and complex surfaces, and Varian Associates Inc. of Palo Alto, California, modified a chip-making machine that can lay down superconducting materials on chips a crystal at a time.

After two years of escalating improvements in superconducting materials, there is now less enthusiasm and information being shared among research laboratories. The breaking of the 77 Kelvin barrier has ushered in a new phase:

the drive to create the technology that will make product possible and the ultimate move to take that product to market. Superconductivity was already a multibillion-dollar business before 1986, and predictions were then that it might reach as high as $36 billion annually by the year 2000. While physicists study superconducting materials and continue to search for the room-temperature ideal, engineers have been pushed to the limit designing technology that will make high-temperature conductors practical. Companies are sobered by the recognition that rapid advances may make products now being developed obsolete by the time they reach the market.

Nevertheless, there seems to be wide agreement among the major participants that superconductor technology will become commercially viable much sooner than the time needed for earlier innovations. Nearly all the experts say superconductivity will make its impact felt before the mid-1990s. Their optimism is based on the fact that much has been known about it for decades and commercial plans were being developed during the 1960s when it was thought that viable superconductors were imminent. Some of the old prototypes and related technology developed then are now being reexamined and updated to accommodate the new materials. The ability to "dust off" and adapt existing technology for today's materials partially explains the optimistic predictions, and some feel the United States is in particularly good shape in the superconductivity race because it leads in the ceramics industry.

Following the key 1986 breakthrough in an American-owned laboratory in Switzerland, U.S. researchers appear to have recaptured the lead in new advances in superconductivity. But will America be able to transport its scientific expertise to the marketplace in the form of tech-

nology and product that will create prosperous new industries and perhaps millions of new jobs? The economic stakes
are incredibly high and the competition will be fierce. The
Japanese have clearly marked superconductivity as an area
of national focus and the Ministry of International Trade
and Industry has poured enormous amounts of money into
cooperative research projects involving academic, industrial, and governmental laboratories. The U.S. government
doubled its superconductivity budget through 1988, but the
bulk of the American effort has been supported by private
funding.

It's too early to predict a national winner in the race
to harness the potential of superconductivity, but the winner's spoils will be considerable and the eventual benefits
to society incalculable.

"SUPER SPIN-OFFS" FOR THE 1990s

Each case history in this book demonstrates a rather remarkable truth: Genuine innovation inspires imitation and
a host of other legitimate innovations. It remains a mystery
to most outside the scientific community, but superconductivity has created enormous excitement within it because of
its almost unparalleled potential. It could prove to be one
of the greatest scientific breakthroughs ever because of the
strongly positive effect it figures to have on so many of the
innovations that preceded it.

There is virtually no end to the wonders it is capable
of producing before the close of the twentieth century, although much work remains to be done. Nevertheless, it is
instructive to look at developments taking shape in just a
few general areas where its impact will be felt.

Computers. Bell Labs created its first Josephson Junction device in 1963, based on the superconducting material available at the time. It proved its ability as an amazingly fast on-and-off switch, and, with today's new high-temperature superconductors, it will be an integral part of the superfast computers that seem destined to appear soon. The speed and size of today's computers are limited by the designers' ability to place only a limited supply of circuits into chips and those chips into the machine. Because electrical resistance creates heat, the space is finite. Superconductivity circuits inside tomorrow's computers can be jammed into tighter areas without the risk of overheating. The shorter the wires that carry the signals the faster and smaller the computer.

Medicine. About 80 percent of today's superconducting wire is used in magnetic resonance imaging (MRI) machines. That wire is forced to rely on liquid helium cooling, making both the machines and their use very expensive. The machines are large—6 feet by 8 feet by 10 feet—in part because of the $100,000 worth of insulation needed to protect the liquid helium. Another $30,000 has to be spent annually to keep the superconductive magnets cooled. New materials cooled by liquid nitrogen will drastically lower those costs and enable many more medical centers to purchase a new generation of MRI machines. Similar help is promised for a conventional superconductor device marketed by Biomagnetic Technologies Inc. of San Diego. It can detect and focus on tiny magnetic fields induced by electrical activity in the brain such as the discharges that accompany epileptic seizures. These high-tech tools may enable surgeons to pinpoint and remove portions of errant brain tissue that create the seizures.

Electric power. Reduced bills from the electric company would seem to be the most obvious benefit the individual consumer might derive from superconductivity. Metal alloy wires cooled by liquid helium have already proved to be far more efficient in carrying electrical power, and, when new superconducting lines cooled by liquid nitrogen are perfected, the utility companies could save huge amounts of money by switching over to the new technology. Replacing the existing systems would require a monumental investment, but obtaining rights-of-way for overhead high-tension lines has become an increasing problem for utilities, and installation of underground superconductive lines may be an attractive alternative.

In the long term, utility companies could save as much as 15 percent of the electrical power that is generated today and lost through resistance in the lines that carry it. In an early experiment, abandoned because of federal budget cutbacks, Long Island's Brookhaven National Laboratory experimented with superconductive cables that were able to carry an eighth of New York City's electrical needs in a pipe just 16 inches across. Superconductivity will permit electrical generators that are smaller, cheaper, and twice as powerful, and scientists are now working on devices that will actually permit the storing of electrical power. Normal technological advances of recent years have allowed cities to request and receive electrical power from distant areas to prevent blackouts, but some parts of the electric net are not so lucky and the threat of power failures remains a problem. A magnetic field is created when current passes through any conductor, and technicians are trying to devise superconducting coils from the ceramic materials that will permit the storing of a huge magnetic field that can be tapped whenever power is needed. Theoretically, the contained current will last forever when the loop is closed.

Transportation. With the skies becoming increasingly unfriendly because of heavy traffic above major airports, passenger trains capable of moving nearly as fast as prejet

airliners will be a welcome innovation from superconductivity. These so-called *maglevs*—they operate on magnetic
levitation—will make Japan's noted bullet train (top speed
149 miles per hour [mph]) and France's famed TGV (186
mph) seem like the steam locomotives of grandfather's day.
Maglev trains are capable of speeds around 300 miles per
hour as they ride smoothly and quietly on the cushion of an
electromagnetic wave. Maglevs have gone past the drawing-
board stage, and prototypes are already being test-run in
West Germany and Japan. The United States dropped its
maglev study in 1975.

Convinced by the mid-1970s that no materials would
be found that could superconduct at temperatures above 77
Kelvin, the West Germans began developing their Transrapid maglev program with conventional magnets. Several
ambitious applications are in the planning stage, including
a 95-mile line between Hamburg and Hanover. But the real
shocker is the German proposal to build a 230-mile link
between Los Angeles and Las Vegas. If a U.S. commission
grants approval in 1989, a Japanese firm, C. Itoh & Co., is
prepared to finance the construction that will enable West
German maglevs to begin running on American soil in the
1990s! Although the German and Japanese systems operate
on similar principles, trains being developed by the Japan
Railways Group and its partners use superconductor magnets cooled by liquid helium. Japanese superconductivity
research and development are thought to be about on a par
with America's, and Japan believes its high-speed trains will
be riding on the wave of the future, literally, when it develops magnets that can be cooled with liquid nitrogen. West
Germany has already spent more than $830 million in public
funds to develop its Transrapid program, and Japan has
invested some $379 million of public and private finances
on its system. The Japanese are taking their usual patient,

long-term approach to new technology and seem confident their system will be the one that will eventually dominate the marketplace.

Some feel that superconductivity will permit the realization of the electric-powered automobile that was dreamed about during the oil crunch of the 1970s. Cars powered by small efficient electric motors that draw their energy from long-lived superconducting batteries could do much to alleviate pollution and reduce the continuing reliance on fossil fuels. Superconductive technology may also propel oceangoing craft. The U.S. Navy has begun to investigate its possible uses, but the Japanese appear to be leading in that area, too. Scientists at Japan's Mercantile Marine University, using a helium-cooled superconducting magnet, have already built a working scale model of a ship that cuts through seawater by electromagnetic propulsion. The designer predicts that new superconductor materials will enable him to launch a 500-ton version of his prototype by 1991.

It may be close to the year 2000 before innovators still unknown are able to use room-temperature superconductors to produce quantities of new devices that will make our current options seem like antiques, but the prospects promise to make the 1990s exciting.

FUSION ENERGY

Scientists continue to labor on a decades-old project that most concede will not become reality in their own lifetime: They hope to discover a practical method of making fusion energy, the same kind of power that exists in our sun and all the stars. The struggle to create that endless supply of energy occupies some of the most innovative minds in the scientific world despite a general feeling that "Fusion always seems to be thirty years away no matter what year it is."

The serious search for fusion energy began forty years ago, and scientists have made steady progress, but the problems they face are so complex that the feasibility of fusion energy has not yet been satisfactorily demonstrated in the laboratory. This has led to pessimistic estimates that the first practical fusion energy plant will not operate until, possibly, the mid-2000s. Fusion energy is viewed as an ideal replacement for current systems that are exhausting the world's diminishing supply of fossil fuels, and that belief has kept the program going despite agonizingly slow progress. Fusion power plants would offer numerous advantages over the systems that produce energy from nuclear fission. Nuclear power plants, as the Chernobyl incident demonstrated, are capable of spreading radiation death through accidents and their growing heaps of spent radioactive fuel. Fusion energy, theoretically, will produce power more effectively and economically with considerably less risk than nuclear power plants. The inner walls of the fusion reactor would become somewhat radioactive after thirty years of being bombarded by neutrons, but scientists insist the reaction chamber could then be dismantled and stored as low-level radioactive waste for thirty to fifty years, when it would become harmless.

Fission and fusion work in similar ways. In both reactions, particles called neutrons fly off the basic materials at high speed inside the reactor. These energetic neutrons are trapped, and the force of their motion is converted into heat that drives turbines to produce electricity. The all-important differences lie in the materials they use. In fission, the nuclei of heavy elements such as uranium split into fragments that are highly radioactive, whereas the nuclei of light deuterium and tritium join together in the fusion process to create the heavier but harmless helium gas. In addition to the extra safety of the fusion reactor, the material that fuels it is cheap and in virtually unlimited supply. Deuterium atoms can easily be extracted from water, and tritium can be produced from lithium, which is also abundantly available.

Unfathomable problems of a physical nature have been largely responsible for science's inability to demonstrate fusion energy. An example is the search for some kind of container that can hold a gas that has been heated to a temperature of 100 million degrees Fahrenheit. This is a problem scientists have labored over for years because the atoms that hold the deuterium and tritium nuclei must be heated into a "plasma" at that incredible temperature in order to induce the nuclei to move about at speeds of 300 miles per second. When the deuterium and tritium nuclei collide at that speed, they fuse and release their incredible energy. Obviously, any material known to man would melt or otherwise disintegrate long before it reached such a temperature.

Scientists are exploring two methods that may overcome that conundrum. An innovative process being examined in several countries bypasses the normal physical walls and relies on an invisible container to keep the superheated plasma confined. The process uses very expensive electromagnets to create powerful magnetic fields that trap the charged plasma in the center of a reaction vessel. To this point, only conventional electromagnets powerful enough to create the desirable magnetic field have been used, and their cost has been prohibitive. Some say fusion energy will move a huge step closer to realization when powerful high-temperature superconducting magnets become available. A totally different experimental method called *inertial confinement* uses powerful laser beams to heat the outer surface of minuscule pellets of deuterium-tritium fuel, causing the rim to explode and the inner part of the plasma to implode. This experimentation has already attracted the interest of the military, allowing them to further their practical knowledge of nuclear weapons physics by studying the miniature eruptions that resemble the explosion of H-bombs.

Government funding for America's research on fusion energy will also be a major deciding factor in how soon it may be a practical alternative to the methods currently in use. The Department of Energy's fusion budget reached an all-time high in 1977 during the height of the oil crisis, but the investment in fusion research has been drastically reduced in recent years. A partial explanation is the government's desire to spread appropriations around to several scientific projects it considers vital, including the superconducting super collider (SSC). Twenty-two states spent more than $30 million trying to win approval as the site of the estimated $5.3 billion SSC project, but the Energy Department gave Texas the nod in November 1988.

Unlike those of the seemingly imminent superconductivity and the long-range fusion energy, the benefits of the superconducting super collider are nebulous. The area in which it is located will immediately benefit from thousands of construction jobs and from another four thousand to seven thousand scientific and technical positions when it is ready for operation, but the SSC's function will be pure research designed to give scientists insights into the most basic nature of matter that might lead ultimately to unimagined benefits for humanity.

The SSC plans a buried ring, fifty-three miles in circumference. It will use a magnetic field created by powerful superconducting magnets to confine proton beams that will travel at nearly the speed of light, in opposite directions, until they collide with an energy of 40 trillion volts. The debris from that collision will provide scientists with copious information never before available. The American super collider will perform at an energy level twenty times stronger than anything that preceded it, meaning that scientists should be able to comprehend at least twenty times more detail about matter and the universe than was ever possible in the past. Some have described the super collider as the largest commitment to basic research in history.

As in the case of fusion energy, the arrival of a new generation of superconducting magnets from high-temperature ceramics would be a boon to the SSC. The current design will rely on some ten thousand of the available low-temperature superconducting batteries cooled by costly liquid helium to keep the protons on course as they speed around the mammoth oval. The expense of those batteries and their complicated refrigeration systems account for a large portion of the $5.3 billion budget, and some scientists have argued that the project should be delayed until new superconducting batteries that will be able to do the job less expensively and more effectively come into existence. Dr. James Krumhansi of Cornell, president of the American Physical Society, was an early advocate of allowing more time, believing that the new high-temperature superconductors might permit the radius of the super collider ring to be reduced to as little as ten miles.

The super collider is scheduled to be completed in 1996, but few expect sensational revelations from this or fusion energy research before the end of the century, at the earliest.

GENE MAPPING

Biotechnology—the ability to clone in a laboratory the natural proteins that are the regulators of the human body—

has emerged since the discovery of the structure of DNA in 1953 as a towering scientific innovation that has revolutionized the life sciences and evinces a capacity for producing more weapons against disease and disability before the turn of the century. Work on a remarkable tool that will further that goal, a genetic map, has been proceeding for some time, and many are calling for an all-out national effort to bring it to fruition.

Scientists in several research centers have been working on intricate maps that help pinpoint human genes, a task so complex it boggles the minds of some within the biomedical research community. How far this research has progressed remains a subject of controversy, but a committee of America's National Academy of Sciences (NAS) has recommended that the federal government appropriate $200 million per year for fifteen years to produce a map of the genome. Molecular biologists say the human body contains 3 billion base pairs of deoxyribonucleic acid or DNA, which means it would cost the taxpayers about $1 to locate each pair for a total expenditure of $3 billion. Critics of the NAS proposal say the size of that expenditure would cramp other research that offers more immediate dividends, but its supporters argue that a detailed genetic map will be an invaluable aid that will direct researchers to cures for many genetic diseases.

A simple rundown of the basic manner in which genes work in the body will help explain the monumental task facing the genetic cartographers. Genes, passed on to us by our parents, are stored on molecules of DNA in base pairs that form a distinctive double helix. They carry specific codes that describe the proteins our bodies need to function. The order of these hundreds of millions of DNA base pairs is part of the code that determines the protein the gene is to produce. That means the human genome contains between

50,000 and 100,000 genes that are stored on twenty-three pairs of chromosomes, in the form of double strands of DNA, ingeniously folded to allow them to fit in their designated area. Of course, none of this complex machinery is visible to the naked eye.

In more common terms, as richly detailed city maps allow tourists to find specific addresses along the narrow, twisting streets of lower Manhattan or Boston's Beacon Hill, a genome map will permit molecular biologists to place "markers" that locate genes, investigate linkage, and discover the types of proteins they are programmed to produce, and make it easier for biotechnology to create monoclonal antibodies that have the power to preserve human life. Genes act as hereditary blueprints of the human body, and the biomedical sciences will be better equipped to deal with aberrations in human development when more knowledge is gained.

Considerable progress in genetic mapping had been made even before the NAS proposal for a fifteen-year crash program to create a sequence of the entire genome. Individual research teams have been forced to create their own maps in order to locate a specific gene, and much larger genetic maps have already been hailed as great achievements. One such map, a joint effort by Massachusetts biotechnology researchers from Collaborative Research, the Whitehead Institute for Biomedical Research, and the Massachusetts Institute of Technology, was detailed in the October 1987 issue of the scientific journal *Cell*. One of the authors of the study, Dr. David Botstein of MIT, called the development a milestone that would permit the linking "all over the genome" of genes that cause diseases. After four years of work, the map was said to be about 95 percent complete. Yet another map, developed independently by Dr. Ray White of the University of Utah, was reportedly at a comparable stage of development near the end of 1987.

Experts say these maps, not nearly as detailed as the one proposed by the NAS, should be enormously helpful in identifying the genetic components of high-risk problems such as heart disease and cancer that may result from two or more genes. The maps, using techniques developed by White and Botstein, mark restricting fragment-length polymorphisms that are each found in only one place in the entire human genetic structure. It is believed that these identifiable genetic variations called polymorphisms can be traced through families to demonstrate patterns of inheritance. Some three thousand diseases, including some of the most deadly, are known to be inherited. The genetic map made by Utah's Dr. White pinpointed the gene responsible for familial adenomatous polyposis, which can lead to early cancer of the bowel. Markers on these existing maps have also discovered the genes for Duchenne muscular dystrophy, retinoblastoma (an eye cancer in children), and Huntington's disease.

More than four hundred such polymorphism markers are now in place on genetic maps developed by White, Botstein, and others, meaning a marker already exists for about one in every 10 million base pairs of DNA. This is no small accomplishment in itself, and it could lead to the discovery of new information that may permit the gene-splicers to develop unimagined remedies for the treatment of cancer and other terrible afflictions. But those who are pushing for a national commitment to develop a survey of all 3 billion base pairs of DNA say it is not enough. Their view is that the information obtained to this point is sketchy at best. They want federal financing for the organization and implementation of a vast project that will utilize teams to research each predetermined section of the genome, with subsequent meticulous review of each completed section, until the whole genome is sequenced.

There is no unanimity within the scientific community favoring the project. And the estimated $3 billion price tag, on top of the $300 million government agencies are already spending on related genetic research, is only one of the objections raised. Many say the mapping of the entire genome is unnecessary because only about 2 percent of it contains information that might be genuinely helpful. Experience has proved that an investigation of five thousand base pairs will usually provide the description of a single protein. It is believed that the genes produce a total of something less than 100,000 proteins, indicating that approximately five hundred pairs contain all the really valuable information to be had. Critics say the other 98 percent of the genome is basically superfluous, making the big study time-consuming and wasteful.

Even the supporters of the NAS plan agree with some of this criticism, but they can point to innumerable scientific discoveries made from seemingly irrelevant material. They also admit that advances are being made without a sequence of all the DNA base pairs but are convinced that a valuable fringe benefit from a commitment to genome sequencing will be the perfecting of innovative technology—fast data-finder chips, neural computers, and so on—that will simplify that project and all other DNA research.

Much of the information the NAS committee's genome sequencing project is expected to provide will not be known until well after the turn of the century. Fortunately, other independent scientists are continuing to map genes, and DNA studies will go on providing extremely valuable information about the physical makeup of man. In only a few decades biotechnology has demonstrated an ability to produce significant health-care products, and the arrival of new genetically engineered proteins, hormones, enzymes, and vaccines in the remaining decade of the twentieth century seems inevitable.

PROSPECTS AND PROBLEMS

The mammoth projects reviewed to this point carry the promise of widespread improvement in fulfilling many of society's basic needs, but there are countless other innovations, too numerous for confinement within a single chapter, that will also brighten the future. Some might be considered frivolous in comparison to those already described, but innovation that can be translated into successful product demonstrably affects the economies of the world. That, in itself, is an awesomely important consideration for the future everywhere, particularly in the United States.

In the past forty years or so American innovators have consistently demonstrated a remarkable talent for discovery—the ability to combine dedicated research and creative thinking and conceive completely new methods of doing things—in such areas as electronics, communications, materials, and biotechnology. There is strong evidence to support the thesis that other countries are rapidly closing America's lead in productive research. But the familiar plaint that America's researchers have somehow lost the knack for innovative thinking is simply not substantiated by fact. It is perhaps more reasonable to look at related areas when pondering why so many of the things in the overflowing horn of plenty, one symbol of America's wealth and well-being, now wear "Made Elsewhere" labels.

Time and again American scientists have created remarkable innovations—that have led to revolutionary new

technologies—that have produced successful products—
that have ended up being exploited in a masterly way by
industries—in countries other than the United States. This
simplistic, generalized rundown begs an answer to a straightforward question: Does American industry still know how
to bring a product to market? Available evidence indicates
that marketing may well be that part of the "Yankee ingenuity" equation the United States has lost. As America
looks toward the future, leaders in the private sector must
encourage the development of vigorous new techniques that
will allow their own firms and the U.S. economy to make
better use of the innovators' creations.

Recent history is packed with examples of America's
inability or unwillingness to compete for huge markets it
appeared capable of dominating, and there are signs those
problems will worsen before they improve. The departed
VCR industry has become a classic illustration of squan-
dered opportunity. Yet the current rush toward high-defi-
nition TV in the rest of the world presages another $25 to
$50 billion industry that U.S. companies seem disinclined
to enter. France and England now appear to be enjoying
some admittedly expensive success with their supersonic air-
liner programs—the United States dropped SST research
years ago. West Germany and Japan are investing hundreds
of millions of dollars for maglev train systems—America
seems a likely customer because it no longer has plans for
its own.

Apparently stung by criticism that it was lagging behind the United States and others in actual innovative research, Japan's Science and Technology Agency instituted
in 1985 what it calls the Frontiers Research Program, "that
will serve as the nucleus of the twenty-first century's science
and technology." And, with the Common Market nations
scheduled to take a giant step toward European unity on

January 1, 1992, even stronger economic competition can be expected from that quarter. Practically all trade barriers between the European member countries will be dismantled, inevitably expanding the marketplace and engendering additional resources for public and private research.

In the autumn of 1986 the Congress of the United States was facing the unenviable task of dividing a limited amount of federal funding among six megaprograms of scientific research: the superconducting super collider at an estimated cost of $5.3 billion, the Hubble space telescope that is expected to rise into the same price range, the human genome map with a price tag of $3 billion, a project to study changes in the global environment that will require an investment of $2 billion, a $13 billion manned space station, production of a hypersonic space plane with an estimated budget of more than $3 billion, and the Strategic Defense Initiative, with estimated expenditures so high that not even an acceptable "ballpark figure" has been able to stand the test of brevity.

Assuming that these programs are equally worthwhile within a broad spectrum of values, can the country afford to pay for them all when so many other standard services require federal funding? A number of national economies have gained glowing health because their governments either funded or helped fund practical research and development programs. In many instances these programs led to commercial applications that play a familiar role in American life. The U.S. government has not been tightfisted in appropriating money for research, but, from a practical business viewpoint, its record for obtaining an appropriate return on its investment is questionable. Clearly, some better mechanism than the systems now in place must be found to decide priorities—on a governmental level—that will permit

American innovators to create the products that will benefit their fellow citizens and all of mankind.

There are many innovative wonders on and beyond the horizon that will make our futures better because of imagination, dedication, and expertise of individuals and organizations all over the globe. A rededication to the can-do spirit of the past and the discovery of an effective mechanism for deciding scientific priorities, in balance with humanitarian and economic needs, would help America's innovators to make many more exciting contributions in the innovative years ahead.

Afterword

Organizations inclined to pay only lip service to innovation—regarding it as a lofty-sounding idea bound to impress the stockholders—must remember its direct relationship with newness and be forewarned that an attempt to gain financial reward through its use is going to be anything but an easy short-term project. There is no shortage of evidence to support the theory that a genuine dedication to innovative approaches in all areas has become an imperative for the future. The old axiom "Why fix it if it ain't broke?" no longer has much relevance. A better slogan for a rapidly changing era might be "Improve it or they'll remove it!"

In recent years we have advanced from the vacuum tube to the transistor to the microchip, from typewriter carbon paper to the modern duplicating machine, from the early death certificate to organ transplants, miracle drugs, and genetically engineered health aids—because forward-looking, determined people were convinced there was a better way. But innovation does not occur in a vacuum, and a look back at the key points raised in this book seems appropriate.

Great innovations often spring from the focused thinking of an intelligent, imaginative individual—a phenomenon that is as prevalent today as it was in the past. The most noticeable difference between the innovators of the past and those of today is that the latter have much better access to a larger body of knowledge. This, in itself, is an indication of the innovative processes that have hastened the arrival of the communications age and the high-tech revolution. Trial and error is still an integral part of scientific research, but newer approaches have often reduced the discovery time span. Chester Carlson, inventor of the xerography process, is the only one of our innovators who fits the classic mold of the solitary genius who struggles alone. It is no accident that his travail began in the earliest time period of our investigations and that he was forced to turn to others before his dream began to be fulfilled in 1950. Yet the key to making xerography a commercial success was an organization that allowed extensive innovation in development, manufacturing, and marketing of his invention.

Another of our innovators, Fred Smith of Federal Express, was an entrepreneur rather than an inventor. He also displayed enormous individual zeal in launching the innovation that was the product of his own mind, though he is the first to insist that he didn't do it alone. Far more common in recent years are brilliant individuals such as Brattain, Bardeen, Shockley, Maiman, Alberts, Borel, Maurer, and Ginsburg, whose achievements resulted from work performed in commercial laboratories and research centers.

Significantly, few if any of these industrial innovators had personal financial gain as the primary motivation for their efforts. The quest for knowledge, the desire to improve the lot of humanity, and the prestige of historic accomplishment were of primary importance to most of them. Of course, it must be stressed that few had much choice in the matter. Although permitting the name of the individual scientist to be given on the patent, virtually all major research organizations require employees to sign advance agreements that automatically confer ownership of the patents—and the subsequent financial rewards—to the employer. This has always been a contentious issue between scientist and corporation, and it is no coincidence that many of the innovators discussed in this book left their parent companies soon after making their discoveries in an attempt to form their own businesses and thus profit from their work.

Some inventors say they want to see the issue of patent ownership decided in the courts, and there is much to support both sides of the argument. It seems self-evident that a person should be entitled to the maximum return for the fruits of his own endeavor. On the other hand, as the corporations argue, who is to say the individual would have been able to make his breakthrough without the wide range of security offered by the parent company in paid time and scientific and financial support? History must not obscure the roles of the special individuals who proved the rightness of their theories, sometimes with the halfhearted support of management, but there is no doubt that most recent outstanding innovations emerged from collective efforts in programs sponsored by organizations. This verifies the value of innovative research and is reason enough for business leaders to examine the wide potential of innovation within their own companies.

The cases reported in this book repeatedly demonstrate that commercially successful innovation must be accepted as a long-term process that requires a variety of total commitments from organizations that hope to capitalize from it. Consider the time spent by numerous researchers at Bell Labs in investigating the photovoltaic effect in silicon until the three inventors produced a working transistor, then add the years that passed before others elsewhere began to find important applications for the device. Theodore Maiman took approximately two years to develop the first working laser after Townes and Schawlow had published their breakthrough paper on the subject, but finding commercial uses for that and the many other types of lasers that followed is an ongoing process that seems destined to produce even more important applications in the years ahead.

Innovators at 3M labored for eleven full years before they could produce any kind of commercial product out of the mysterious and risky fluorochemistry project the company had started in 1945. As our studies of Sandoz and Merck demonstrated, the time needed to discover a miracle drug is often measured in decades, and the firm that pays for the research cannot expect to begin recouping its investment until long after the breakthrough is achieved. In every instance, our examinations show that the potential rewards of innovation must be carefully weighed against the time, expense, and possible failure of the entire process.

Can innovation succeed only within the framework of a small organization? We feel a number of stories in this book disprove this widely held theory, although we cannot argue with the contention that the innovation process is often easier in a smaller organization. Because it requires deep commitment, cooperation, and close control, innovation flourishes best in an environment where it is the centerpiece rather than a sideline. An urgent need to come up with marketable new products forced small companies in California and New York to begin developing audio- and video-tape recorders and experimental machines that might produce dry copies of documents. The case histories of those companies, Ampex and Xerox, demonstrate the ability of innovative planning and action to overcome all obstacles in creating new products capable of transforming small companies into major corporations. Their success inspired the start-up of entire new industries, but both Ampex and Xerox were eventually victimized by a rapid-growth syndrome that necessitated a painful reconsideration of goals and methodologies.

Do successful, long-established companies have too much to lose by rocking the boat with risky adventures in innovation? The visible wilting of once-invincible U.S. industrial giants under the pressure of innovative foreign competition—coupled with the accomplishments of Bell Labs, Merck, Corning Glass, and others—clearly indicates that that business commandment is no longer carved in stone. Bell Labs, one of the biggest and best research organizations in the world, turned out important advances before and after the transistor because of its established program of encouraging innovation. Merck occupied a similar position among pharmaceutical companies, but it dared to invest both time and money when it brought in Roy Vagelos to institute fresh approaches to new drug discovery. Under Vagelos's direction new labs were set up, methods devised, and researchers brought in that eventually paid strong dividends, elevating Merck to the rank of most-admired corporation in America.

Corning, an old-line American business with a solid history of success in the production of household glass products, also opted to expand its horizons through the innovative skills of its own scientists. Management, in an admittedly cautious original commitment, offered a group of its researchers the challenge of using their knowledge of glassmaking to create something that had eluded telecommunications scientists for decades. That conservative unleashing of its scientific innovators brought such rapid and astonishing results that Corning must be credited with the key breakthroughs that opened up the age of fiber optics, which is still in its infancy.

To find the pluperfect example of how genuine innovation can prosper in a huge company one needs to look no further than Minnesota's 3M Corporation. Here is a one-time mining company that becomes ever bigger and more successful because of its continuing dedication to innovation. Their formal recognition that "people have a basic urge to create" has led 3M's managers to devise strategies that foster internal communication and shared information and provide employees with challenge, responsibility, resources, rewards, and sponsorship that keep the innovation process healthy. These strategies have resulted in a wide variety of products that are spun off into new corporate divisions, specifically designed to avoid the pitfalls of bigness.

Because of innovation's obvious relationship to newness, it would seem to follow that companies that expect to profit from its application must be pioneers. Our research demonstrates that this is not always the case. The misconception probably lies in the subtle differences between the words *invention* and *innovation*. A company that introduces a new invention is not always capable of finding ways to make immediate and marketable use of its potential. Such was the case with Bell Labs and the transistor, as well as Hughes Aircraft and the ruby laser. As others have aptly noted, few in the automobile industry can readily name the man who actually invented the automobile, but most will profess some knowledge of Henry Ford, whose method of mass-producing cars made them available to a large American public. This interplay, between invention and a viable product/service, is where sensitive and understanding management determines fizzle or success. And it is the increasing need for world-class corporate scale to permit success in this transition that presents managers of the innovation process with their most crucial challenge.

Chapter 9 contains an entire series of events that illustrate the fact that pioneers don't always win the monetary sweepstakes. European and Japanese concerns first experimented with the development of cash-dispensing machines. But America's Docutel created the magnetic-stripe technology that made it the leading producer of automatic teller machines and cleared the way for ATMs to become the cornerstone of electronic banking. Other companies then improved on the Docutel machine and achieved even greater success. Chemical Bank pioneered the use of ATMs in the United States while the rival New York Citibank watched and learned before moving boldly with sweeping strategies that proved just how valuable the machines could be to the public and the banking industry. In each instance, one is forced to wonder, which was the innovator—the pioneer or the latecomer?

An even more striking example that victory does not always come to the front-runner is to be found in the history of videotape recording. The engineers at America's Ampex created the technology that cleared the way for the high-quality reproduction of television pictures and sound, and RCA and other U.S. electronics firms added their share of important refinements. But none of them succeeded in reinforcing technological skills with sound business practices—production and marketing—that would permit them to stay the course. Japanese competitors, in a dazzling display of speed and strategy, refined the original technology and defined and captured a market far bigger and more profitable than that of the pioneers.

*   *   *

These and other examples demonstrate that discovery, as vital as it is, is but one phase of the process of innovation that makes it possible for business to prosper and humanity to profit. None of the new concepts or products studied on these pages would have had an opportunity to benefit the general public except for the strategies that brought these innovations to the marketplace.

Clearly, there can be no single foolproof formula for fresh thought and action that will be applicable to every field of endeavor. But there are recurring themes in the case histories in this book that, put together, form a distinct outline for a workable approach. They are discernible attitudes that have given organizations a distinct competitive edge when consciously nurtured and actively practiced on all levels. Companies that profit from innovation must, first of all, be aware of the time and risk factors inherent in that commitment and must be prepared to balance the continuing profits from established products against the financial risk of untested new areas. Discovery does not automatically translate into commercial success, and the rewards tend to be commensurate with the level of risk. Virtually every company that has profited from innovation found its own approach only after continual reexamination, debate, and testing.

Innovation invariably breeds other innovations, and it is this capacity for proliferation that has led us into this era of rapid change since the end of World War II. It is both encouraging and important to note at this point in our history—when there seems to be such an emphasis on *things*—that the more successful and innovative a company is, the higher the premium it places on human resources. Genuinely innovative companies recognize the value of human talent and treat it as capital: a theme as old as yesterday's hopes and as new as tomorrow's aspirations, which is as it should be. The collected knowledge of the past has made it easier for today's researchers in their quest for new discoveries.

In approaching the topic of innovation by examining the process as well as the product, we are left with few reservations about its continuing potential as an economic method or device. But innovation cannot begin and end in the laboratory or research center. It requires the coordinated efforts of many process innovators to effect real change and to shepherd a new product or service to market. Innovators have been active since the first prehistoric man learned to use fire for his tribe's benefit. Their *worth* has increased as they participated in the shrinking of the planet and the enhancement of life throughout the world. So it seems only fair to say their *work* will not diminish in importance for as far into the future as the human mind can comprehend.

Bibliography

BELL TELEPHONE LABORATORIES, TECHNICAL STAFF. *A History of Science and Engineering in the Bell System*. New York: Bell Telephone Laboratories, 1975, 1983.

BERNSTEIN, JEREMY. *Three Degrees Above Zero: Bell Labs in the Information Age*. New York: Charles Scribner's Sons, 1984.

BODE, H. W. *Synergy: Technical Integration and Innovation in the Bell System*. Murray Hill, N.J.: Bell Labs, 1971.

CHANDLER, ALFRED. *The Visible Hand: The Managerial Revolution in American Business*. Cambridge, Mass.: Harvard University Press, 1977.

DIBACCO, THOMAS. *Made in the USA: The History of American Business*. New York: Harper & Row, 1987.

DRUCKER, PETER F. *Innovation and Entrepreneurship: Practices and Principles*. New York: Harper & Row, 1985.

GILDER, GEORGE. *The Spirit of Entrepreneurialism*. New York: Simon & Schuster, 1984.

GRAHAM, MARGARET B. W. *RCA and the Videodisc: The Business of Research*. New York: Cambridge University Press, 1986.

JACOBSON, GARY. *Xerox: American Samurai*. New York: Macmillan, 1986.

KAY, NEIL M. *The Innovating Firm: A Behavorial Theory of R&D*. New York: St. Martin's Press, 1979.

THE INNOVATORS

KETTERINGHAM, JOHN, and NAYAK, RANGANATH. *Breakthroughs*. New York: Rawson Associates/Macmillan, 1987.

LAMBRO, DONALD. *Land of Opportunity: The Entrepreneurial Spirit in America*. Boston: Little, Brown, 1987.

MINNESOTA MINING & MANUFACTURING COMPANY. *Our History So Far: Notes from the First 75 Years of the 3M Company*. St. Paul, Minn.: 3M Company, 1977.

MUELLER, ROBERT K. *The Innovation Ethic*. New York: AMACOM, 1971.

STEELE, LOWELL. *Innovation in Big Business*. New York: Elsevier, 1975.

WALL STREET JOURNAL STAFF. *The Innovators: How Today's Inventors Shape Life Tomorrow*. Princeton, N.J.: Dow Jones Books, 1969.

WALTON, RICHARD E. *Innovating to Compete: Lessons for Diffusing and Managing Change in the Workplace*. San Francisco: Jossey-Bass, 1987.

Acknowledgments

My special thanks go to Charles Francisco whose research contributed greatly to my book.

The individuals whose innovations are portrayed in the various cases, together with their loyal staffs and colleagues, have also played significant roles. I should like in particular to acknowledge:

Nobel Laureate Arno Penzias, vice president of research, Bell Telephone Laboratories, Inc.; James D. Ebert, director, Chesapeake Bay Institute, The Johns Hopkins University; Professor Andrall E. Pearson, Graduate School of Business Administration, Harvard University; Nobel Laureate Dr. William Shockley; James P. McMahon, public relations director, AT&T Technology Systems; Richard Muldoon, Media Relations, Bell Laboratories; Fred Smith, chairman, Federal Express Co.; Shirlee Finley, Fedex Public Relations; Dr. Max Link, member, Executive Committee, Sandoz Ltd.; Dr. Jean Borel, Preclinical Research, Sandoz, Ltd.; Craig Burrell, M.D., Sandoz Corporation; Joyce Yaeger, Brown

& Powers Associates; Allen F. Jacobson, chairman and chief executive officer, 3M Company; Lester C. Krogh, vice president, Research and Development, 3M Company; Mike Nelson and Leon Carr, Public Relations, 3M Company; William Glavin, president, Babson College, formerly vice chairman, Xerox Corporation; Edward Finein, vice president, Product Delivery Processes, Xerox Corporation; Thomas G. Crosby, editorial assistant, U.S. Marketing Group, Xerox Corporation; John Rasor, director, Public Relations, Xerox Corporation; Dr. Allen E. Puckett, chairman emeritus, Hughes Aircraft Company; Dr. Arthur N. Chester, vice president and director, Hughes Research Labs; Dr. Theodore H. Maiman; William Hermann, director, Public Relations, Hughes Aircraft Company; Nobel Laureate Dr. Charles H. Townes, University Professor Emeritus, Department of Physics, University of California, Berkeley; Susan J. Klinger, Business PR Supervisor, Corporate Communications, Corning Glass Works; Dr. P. Roy Vagelos, chairman, Merck & Company, Inc.; Al Alberts; Dr. Arthur Patchett; Dr. Jonathan A. Tobert; Richard C. Bostwick, director, Public Relations; Art Kaufman, manager, Communications PR—all from Merck & Company, Inc.; James W. Johnson, division executive, Citibank; James Kelly, vice president, Chemical Bank; Phil Megna, vice president, Chemical Bank; Thomas R. Metz, vice president, Electronic Data Systems; Don Bartoo, PR, Diebold, Inc.; Mary Ann Jackson, PR, Diebold, Inc.; Gordon Piercy, vice president, Puget Sound Bank; Liam Carmody, president, Carmody & Company; William Hittinger, former executive vice president, Research and Engineering, RCA Corporation; Charles A. Steinberg, executive vice president, Sony Corporation of America; Dr. Michael P. Schulhof, vice chairman, Sony Corporation of America; Grace Ann Yee, International Corporate Communications, Sony Corporation of America; Karen M. Calderone, Corporate Communications, Ampex Corporation; Debra Bannister, director, Corporate Communications, Genentech Inc.

Index

Index

302